SURGERY WITH SPECIAL REFERENCE TO PODIATRY

By

Maximilian Stern

and

Edward Adams

Édition et mise en page : Culturea (Hérault, 34)
Retrouvez notre catalogue sur http://culturea.fr/
Contact : infos@culturea.fr
Imprimé en Allemagne par Books on Demand Gmbh
Design typographique et layout : Derek Murphy et Reedsy
ISBN : 9791041985722
Date de parution : février 2024
Ce livre :
— a été composé avec une police open source libre de droits dessinée sur Open Fondry https://open-foundry.com/

— permet de replanter un arbre. SHS Éditions soutient Plantons Pour l'Avenir, fonds de dotation pour le reboisement des forêts françaises
400 projets de reboisement de forêts soutenus depuis 2014 à travers la France https://www.plantonspourlavenir.fr/

CHAPTER I

INTRODUCTION

Surgery, in contradistinction to medicine, as a separate branch of the healing art, includes all manual procedures and is not limited to cutting operations exclusively. It is that branch which deals with the treatment of morbid conditions by means of manual or instrumental agencies.

Morbid conditions include three distinct classes: those due to

1. Injuries

2. Infections

3. Diseases

Injuries. To this class belong all the processes due to physical agencies and it includes besides traumatism, the effects of heat and cold, of chemicals, of light and of electricity.

Infections. These may be either local or general. The reaction might occur at the point of entrance of the bacteria, or constitutional symptoms may evidence their invasion into the blood, or the absorption of their toxic products.

Many conditions in this class are linked closely with those in the following class:

Diseases. Here are classified (*a*) new growths, both benign and malignant; (*b*) changes due to age and environment, and (*c*) diseases not belonging in either of the above classes. These latter are generally known as *idiopathic* or *spontaneous* in their etiology. These terms, however, often indicate only a limit of knowledge as to their true etiology.

The Tissues. The tissues of the body, though apparently so different and varying so decidedly in their functions, are in many respects similar.

Every tissue is composed of two parts: the *cellular elements* and the *intercellular substance*. These are called *cells* and *stroma*. Upon the first of these depends the vitality and function of the part, while its density, shape and general physical properties are determined by the second. Likewise along the same lines of reason, all of our organs have two separate areas of tissue: the*parenchymatous* and the *interstitial*. The first contains the functioning and the second the supporting elements.

The physical conditions of the interstitial tissue or the intercellular substance vary greatly in density. *Blood* is a tissue, the intercellular constituent of which is fluid, and as we consider more dense tissues, we encounter all degrees of density of the framework or intercellular substance, until with the additional presence of calcareous elements, we conceive the hardness of bone and dentine. Tissues as a whole, however, are not solid; there are spaces in the supporting structure to admit of the passage of arteries, veins, nerves, and lymphatics.

Abnormal conditions arise in the various parts of the tissues. Certain diseases affect the parenchymatous tissue in an organ more than the interstitial tissue and again others affect the blood vessels particularly.

There may be *hypertrophy*, in which the entire organ or part becomes larger as a whole, the active cells and stroma sharing alike in the process, or there may be an *hyperplasia*, in which the active cells of the part proliferate abnormally. When the interstitial tissue alone develops excessively it is known as an *infiltration*. Under such circumstances the parenchymatous cells often undergo what is termed *pressure atrophy*; they are diminished by squeezing.

Atrophy of the part or organ, from whatever source, signifies its diminution in size; its function is, of course, either impaired or suspended as the process goes on.

The efforts on the part of the organism as a whole to combat or repair injury, bacterial invasion and disease are directed along definite lines. The study of these functions does not rightly come within the scope of these pages, belonging to physiology, but must be considered here, if only in brief for the purpose of a good understanding of the processes at work in surgical conditions.

As surgeons, the functions which concern us most are the

1. Nervous

2. Circulatory

3. Lymphatic

THE NERVOUS SYSTEM

The nerves operate in harmony with each other and with the organs to maintain health. The nervous system comprises the brain, the spinal cord, the nerves, and the ganglia. Aside from presiding over the special senses, this system controls and directs the processes of defense and repair. In doing so, the force and frequency of the heart beat, the calibre of the vessels and the chemistry and composition of the blood are all altered.

These phenomena are the ones which concern us in our present subject, acting as they do upon the blood and the organs which contain it. They will be considered more fully in the following:

THE CIRCULATORY SYSTEM

In health and disease the force and frequency of the heart beat and the calibre of the arteries undergo momentary alterations to meet changes surrounding us.

Dilatation and constriction of the arteries, and arterioles through the activity of the vasomotor nerves, permit of increasing and diminishing amounts of blood reaching the various organs and regions of the body. In this way the peripheral circulation, through the activity of the heat centre in the brain, acting upon the superficial capillaries, maintains the normal temperature of the body during the changes of season. In winter, heat conservation is effected by vaso-constriction of the superficial capillaries, while in summer, heat radiation is accomplished by vaso-dilatation together with the evaporation of moisture so abundantly supplied by the active sweat glands. Other phenomena demonstrating the vasomotor function are blushing, going pale, and the redness and swelling following injury or infection. Of the latter we will treat under the heading "Inflammation."

Certain changes also occur in the blood in order that it may perform its functions. These changes are found both in the chemistry of its fluid content and in the number and kind of its solid elements: the corpuscles. The blood is a tissue; its corpuscles are the functioning cells and its fluid content the basement substance.

In the fluid content of the blood or *plasma,* as it is called, certain chemical changes occur in its fibrin-forming capacity. Clot formation, an effort on the part of the economy to arrest hemorrhage, is thus facilitated when there is active bleeding, also during labor and certain diseases.

The number and kind of white blood cells also undergo changes, as we shall see, under circumstances in which the defences of the organism are called into operation, for it is the function of the white blood corpuscles to combat bacterial invasion.

It is the preponderance of these white cells which imparts the peculiar milky color to pus.

Nourishment and oxygen are carried to the tissues by the red blood corpuscles. Under circumstances which we shall see later, these are also altered both in number and in size, and also in their arrangement in the vessels. (See Chapter IV, "*Inflammation.*")

THE LYMPHATIC SYSTEM

Waste material in the tissues is carried off by the return blood in the veins and also by the lymphatics. These are in reality the scavengers of the body, both the lymphatic vessels and glands, performing important functions for the good of the economy. They are found beneath the skin and in the subcutaneous tissue, also along the course of the great veins.

The lymphatics far exceed the veins in number and perform a function similar to the veins, gathering waste material from the tissues, also the digested food called *chyle,* from the stomach and intestines. The lymphatic vessels, also the lacteals which absorb the chyle, all pass through glands, which are extremely numerous, where certain deleterious substances are extracted from the lymph before it is emptied into the blood stream.

The lymphatic glands in performing their function as filters often become inflamed and when large numbers of pathogenic bacteria, or very virulent ones are contained in a gland, abscess formation results just as it would in any tissue so invaded.

CHAPTER II

SURGICAL BACTERIOLOGY

A *microorganism* or *microbe*, some species of which cause all disease, is a minute plant or animal too small, as a rule, to be visible to the naked eye.

The word *germ* may be used to designate any microorganism, but it has so many other meanings and has been so loosely employed even in this sense, that it cannot be used for accurate scientific description.

Bacteria are minute plants on the order of fungae, many of which are able to produce fermentation, decomposition or disease.

Although the word *bacterium* by derivation has the same meaning as *bacillus* and indicates a rod shaped fungus, it has been so loosely employed that it may very well be applied to the entire germal family, retaining the word bacillus in the narrower sense.

Description of Bacteria. *Schizomycetes* is the name given all the bacteria of putrefaction and disease, the former being called *saprophytic* and the latter *pathogenic*.

Bacteria are minute fungi, each consisting of a single cell enclosed in a cell membrane of cellulose which can be demonstrated by iodine, the latter causing the protoplasm to retract from the cell wall. There is no nucleus or central core. Some of the bacteria are colorless, others pigmented, yellow, blue or red. The cells vary in shape and in size in different species as well as in their mode of growth, and are named in accordance with these peculiarities. The round or oval cells are called *cocci;* the rod-shaped organisms are termed *bacilli*. The cocci are called*micrococci* or *macrococci* according to their size; *diplococci* or *tetracocci*, according to the production of pairs or groups of four in their multiplication; *streptococci*, because in their growth they always form chains of cells; *staphylococci*, because they grow in irregular clusters resembling bunches of grapes. Some of the bacteria have the power of motion generally produced by cilia or flagella and others are motionless.

Habitat. These organisms may truly be said to be omnipresent. Every thing we wear or use or eat, even the air itself, is impregnated with them. Pathogenic germs may also be found among these myriads. Every species has its own particular habitat, where the conditions especially favor its growth, just as any of the larger plants require a certain soil, a supply of water, temperature, and proper amount of light in order to make growth and multiplication possible.

The bacteria in the air are more numerous in dry weather, being carried up as dust by the wind, for a moist surface holds any bacteria which may lie upon it. So complete is the action of moisture, that air, which contained 600 microorganisms when inspired, has been shown to return from the lungs with almost none, the moist respiratory surfaces catching and holding the bacteria; so that the expired air is practically sterilized; this is true even when the expiration is from diseased lungs. The act of coughing, however, may expel bacteria in the mucus ejected. The number of bacteria in the air is very variable, but is much greater in houses than out of doors, and is naturally increased by attempts to clean the rooms.

Parasitic Nature. The number of species of pathogenic germs is comparatively small compared with the number of all the varieties of germs, for the latter are practically innumerable. Indeed, the wonderful qualities of resistance in animal tissues is the only thing that makes animal life possible and it is this power of resistance that allows certain wounds to heal by primary union when left without protection or care.

The schizomycetes are unable to extract nitrogen from the air or the soil, like the higher vegetables, and must, therefore, be provided with a higher nitrogenous compound, such as is produced by vegetable and animal life. Some of them are able to live upon dead organic matter, while others cannot exist without living tissues to feed upon and are therefore true parasites. There are some which are able to live upon either dead or living tissues and are known as *facultative parasites*, a class which includes a majority of pathogenic germs. Some organisms require albuminous matter, others need carbohydrates; they all require water, carbon, nitrogen, oxygen, and certain inorganic materials, especially lime and potassium. All organisms require water. If dried, no form will multiply, and many forms will die.

The fluids and tissues of the individual may or may not afford a favorable soil for the germs of a disease, or, in the same person afford it at one time, and not at another. Some individuals seem to possess indestructible immunity from, and others are especially prone to, certain contagious diseases. Impairment of health, by alterating some subtle condition of the soil, may make a person liable who previously was exempt.

Effect of Oxygen. Some bacteria need free oxygen; some can live either with or without free oxygen, while others cannot live at all in the presence of free oxygen. Those requiring oxygen are called *aerobic;* those which can live with or without it are called *facultative aerobic;* those which do not live in free oxygen are called *anaerobic.*

Bacteria are very sensitive to temperature, few being able to live in a temperature below 68°F. or 29°C. or above 104°F. or 40°C. The pathogenic varieties thrive best at about the normal temperature of the blood. Direct sunlight retards their growth and may kill them. Freezing renders bacteria motionless and incapable of multiplication, but it does not kill them; they again become active when the temperature is raised. The absurdity of employing cold as a germicide is evident when it is known that a temperature of 200°F. below zero is not fatal to germ life, cell activities by such a temperature only being rendered dormant. The high temperatures are fatal to bacteria, moist heat being more destructive than dry heat, and adult cells are more easily killed than spores. A temperature less than 212°F. will kill many organisms and boiling will kill every pathogenic organism that does not form spores. Some spores are not destroyed after prolonged boiling and some will withstand a temperature of 120°C. As a practical fact, however, boiling water kills in a few minutes all cocci, most bacilli, and all pathogenic spores, though anthrax and tetanus are harder to kill than are the spores of other bacteria.

Under favorable conditions bacteria multiply rapidly, but when conditions are unfavorable, they take on a spore formation and remain in a quiescent state, like the seed of a plant, waiting—it may be years—until proper conditions are present. The spores are protected by such a thick envelope and have such great potential vitality, that it is much more difficult to kill them than the developed bacteria. Certain spores that withstand 212°F. or 100°C., can be killed when fully developed at 130°F. or 55°C.

Toxins. As bacteria grow, certain poisonous chemical substances appear about them. These poisons are produced by them directly, or are formed in the organic matter or tissues in which they live, as the result of their presence. Some of these substances are alkaloidal and are known as *ferments* or *ptomains*. Others are albuminous in nature and are called *toxalbumins*. The ptomains and toxalbumins are exceedingly powerful poisons, producing local necrosis, inflammation and even suppuration, when introduced by themselves and entirely free from living germs, into the tissues of animals. Pathogenic bacteria abstract the lymph from the blood. As the lymph contains elements necessary to the body, such as water, oxygen, albumins, carbohydrates, etc., their loss brings about body-waste and exhaustion from lack of nourishment. Again, bacteria produce a vast number of compounds, some harmless and others highly poisonous.

The symptoms of a microbic disease are largely due to the absorption of poisonous materials from the area of infection. These poisons may be formed in the tissues by the action upon them of the bacteria, or they may be liberated from the bodies of degenerating microbes.

Bacteria secrete and contain ferments like pepsin or trypsin, and as albumoses are formed in the alimentary canal by the action of the digestive ferments upon proteids, sugars, and starches, so microbic albumoses are formed by the action of microbic ferments upon tissues.

The local and general symptoms of these toxins depend upon the particular toxin employed and a large number of these poisons have been isolated and studied. Those of the surgically important pathogenic germs, produce inflammation locally, with general symptoms of fever, chills, cardiac depression, irritation of the kidneys and bowels and cerebral symptoms, such as delirium and coma. The toxalbumins also appear to have the effect of destroying the bacteria to which they owe their origin when they have been produced in large quantity.

Cultivation. Bacteria are cultivated for study in the laboratory in meat extracts, in gelatine, or agar agar (a sort of vegetable gelatine), or raw potato, in blood serum and in other materials. The simplest method of cultivation is in bouillon, sterilized in flasks, with cotton plugs. Another method of studying bacteria is by the inoculation of animals.

Infection. Bacteria gain admission to the living tissues under natural conditions, by penetrating any of the mucous membranes which they can reach, or by entering open wounds. It may be said in general that an intact epidermis is almost a complete protection against infection, and that an intact mucous membrane is a good protection. This difference in vulnerability between the mucous membrane and the skin is important, and is probably due to the cornifaction of the epithelial cells, and to their numerous layers, as well as to the protection afforded by the thick corium. The single layer of soft mucous cells is much more easily penetrated.

Typhoid bacilli and other hostile germs have been actually observed in the urine, in the bile, in the intestinal secretions and in the saliva. The bacteria of typhoid fever and tuberculosis have been found in the milk of nursing mothers.

The local phenomena of inflammation usually follow the introduction of living bacteria into the tissues, and general symptoms of poisoning follow later, when the bacteria, toxins, or ptomains, have entered the circulation. Some bacteria, however, excite no local reaction, but enter the circulation at once. The pyogenic variety, it should be noted, cause the production of pus.

Elimination. Bacteria can be eliminated from the blood in several ways; the kidneys, however, are the organs which carry the burden of most frequently relieving the body of them. Even the sweat glands are supposed to eliminate both bacterial toxins and bacteria.

Resistance Offered by Tissues. The tissues have considerable power of resistance under ordinary circumstances, although the exact sources of this power are not well under stood.*Phagocytosis*—

the power of destruction and removal of bacteria supposedly possessed by the leucocytes emigrating from the blood vessels—explains it in part. It is also accounted for by the germicidal properties of the blood serum.

The resistance of the tissues may in some cases be due to the absence from them of some particular element necessary to the growth of a particular microorganism. This refractoriness varies in every species of animal in its relation to every form of germ. Different individuals of one species also vary in their susceptibility, and even different parts of the body vary in the same individual. The lower animals offer a greater resistance to pyogenic bacteria than do human beings.

Any cause that lowers the vitality by depressing the system, reduces the resistance to bacteria and is therefore apt to favor their growth. Exhausting diseases such as anemia, obesity, alcoholism, diabetes, fatigue, or even exposure to cold, are instances. Germ growth is also favored by the presence of dead, or injured tissues, of blood clots, of foreign substances, and above all, by the presence of some of the substances in which the germ has already been growing at the time of its inoculation, and containing some of its toxins.

Immunity. To be able to resist the invasion of any species of bacteria, one is said to be refractory to or immune against that variety of germs.

Serum therapy is based on the demonstrated fact of immunity, and of the possibility of producing it by injecting the serum of immunized animals. In many infectious diseases, one attack protects an individual for a lifetime and one form of disease may protect against even a more virulent form, as vaccination protects against smallpox. It is a fact that if the serum of an animal which has been rendered immune to a certain disease be injected into a susceptible animal, the same immunity can be produced temporarily in the second animal. Serum therapy proves that the injected serum will not only confer immunity against the infection, but will enable the animal to throw off an already existing infection.

Sterilization. The question how to destroy microorganisms is one of the most important in bacteriology. Exactly how chemical antiseptics act in suspending the growth in living organisms and yet leaving them capable of restoration, is not understood. The explanation is offered that the antiseptics enter into combination with the capsule of the cell and can be freed from it by breaking up this chemical combination. It has always been evident that very minute quantities of germicidal substances, and some substances which are not germicidal, would prevent the growth of bacteria, so that it is not surprising that chemical disinfectants should act in this prolonged inhibitory way. It must be remembered that in operative surgical work, germs which will not develop are, for practical purposes, as good as dead; therefore such results do not invalidate the present methods of sterilization for operations. They naturally stimulate interest in the discovery of better methods of sterilization and especially in the thorough application of the methods upon which we are now depending, in order to obtain the best possible results from them. There are three ways of destroying microorganisms: (1) by deprivation of food and water, (2) by chemicals (including toxins), (3) by heat.

Chemical Antiseptics. For practical disinfection, chemicals and heat need only concern us. The power of these substances is greatly decreased by heat, grease, oil, mucus, and even blood will cover germs with a coating which prevents chemical germicides from reaching them. Among the ordinary germicides, bichloride of mercury, iodin, alcohol and carbolic acid, are of the greatest importance. A source of error in the direct application of these experiments is the fact that many of these chemicals are decomposed or rendered inert, by combinations with the albuminoids of blood and pus, mercuric bichloride being transformed into an indifferent substance and even carbolic acid being altered.

Carbolic Acid is a valuable germicide in the strength of from 1 to 40, to 1 to 20. It is very irritant to tissues and carbolized dressings may be responsible for the sloughing of a wound. It is inert in fatty tissues.

Carbolic acid is readily absorbed, and may thus produce toxic symptoms. One of the early signs of absorption is the appearance of the urine, which may assume a smoky, greenish or blackish hue.

Examination shows a great diminution or entire absence of sulphates, when the acidulated urine is heated with chloride of barium. The urine also contains albumin. The appearance of the urine is an indication that the use of the drug must be discontinued.

Kreolin, a preparation made from coal tar, is a germicide without irritant or toxic effects. It is less powerful than carbolic acid, but acts similarly, and is used in emulsion of a strength of from 1 to 15%. It does not irritate the skin like carbolic acid.

Peroxide of Hydrogen is a most admirable agent for the destruction of pus cocci. It probably destroys the albuminous element upon which the bacteria live, and starves the fungi.

Peroxide is not fatal to tetanus bacilli.

Iodoform is largely used, but it is not a germicide as bacteria will grow upon it. It hinders the development of bacteria and directly antagonizes the toxic products of germ life.

Silver Nitrate is a valuable antiseptic. It exerts an inhibitive action upon the growth of microorganisms, but irritates the tissues.

Formaldehyde has valuable antiseptic properties. Formalin is a 40% solution of the gas in water. Solutions of this strength are very irritant to the tissues, but a 2% solution can be used to disinfect wounds and instruments.

Nucleins, especially protonuclein, possess germicidal powers. Protonuclein is of value in treating areas of infection, particularly when sloughing exists. A great many other antiseptics are used.

Heat. The surest and quickest method of destroying bacteria is by heat. Even the spores succumb to it. Anthrax spores are killed in 2 minutes in boiling water, and the various bacilli and cocci in from 2 to 5 seconds.

When a substance to be sterilized by heat will not bear so high a temperature, the method of fractional sterilization is employed, the fluid to be sterilized being heated to from 140°F. to 175°F. or to from 69°C. to 80°C., for from 15 to 30 minutes every 3 days or 7 days. The theory is that the adult germs are killed by the first heating and that any spores which develop subsequently are destroyed in their adult state at the next heating. The fluid, meanwhile, must be kept at an even temperature which will encourage the development of any spores it may contain. Even anthrax spores may be killed by 167°F. to 185°F., or 75°C. to 80°C., in a one and four-tenths solution of bicarbonate of soda, in from 8 to 20 minutes. Dry heat is not so efficient as moist heat.

The following are the more important bacteria we meet in surgical conditions:

Staphylococcus pyogenes aureus—a microorganism producing yellow pus.

Staphylococcus pyogenes citreus—a microorganism producing lemon-colored pus.

Staphylococcus pyogenes albus—a microorganism producing white pus.

Streptococcus pyogenes—a streptococcus producing pus. (Erysipelas for example).

Micrococcus gonorrhea—bacillus of Neisser, or gonococcus.

Bacillus pyocyaneous—producing a green pus.

Bacillus coli communis—producing intestinal conditions. (Appendicitis for example).

Bacillus typhosis—Eberth's bacillus; producing typhoid fever.

Bacillus tuberculosis—Koch's bacillus; producing tuberculosis.

Bacillus tetani—Nicolaier's bacillus; causing tetanus.

Treponema pallidum, or *spirochaeta pallida* of Hoffman and Schaudin—a protoza causing syphilis.

The first six are known as *pyogenic bacteria,* as they all produce pus; in addition to the above there are many more microorganisms, but from a surgical standpoint those mentioned are the most important.

The *staphylococcus pyogenes* is a spherical coccus of somewhat variable size but averaging about 8 microns; when properly stained it can often be seen to be formed of two separate hemispheres. In pus it is generally found in small heaps containing from two to ten members, but it also occurs singly and in pairs, and even in short chains like the streptococcus, thus rendering diagnosis difficult with the microscope alone. Its cultures are of a yellowish tinge. The aureus type is the most usual cause of abscesses (circumscribed suppurations) and 77% of acute abscesses are due to the staphylococci.

The *staphylococcus pyogenes aureus* is a facultative anaerobic parasite which is widely distributed in nature, and is found in the soil, in the dust of air, in water, in the alimentary canal, under the nails, and in the superficial layers of the skin. It forms the characteristic color only when it grows in air. It is killed in ten minutes by a moist temperature of 58°C. and is instantly killed by boiling water. Carbolic acid (1 to 40) and bichloride of mercury (1 to 2000) are quickly fatal to these cocci.

Staphylococcus pyogenes citreus, the lemon-colored coccus, is found occasionally in acute circumscribed suppurations, but far more rarely than the other two forms. Its pyogenic power is even weaker than that of the albus.

Staphylococcus pyogenes albus, the white coccus, acts like the aureus, but is more feeble in power. When this organism is found upon and in the skin, it is called *staphylococcus epidermis albus,* an organism which is the cause of stitch abscesses.

Streptococcus pyogenes is found in spreading suppurations and in very acute abscesses. About 16% of acute abscesses contain streptococci. It is easily killed by boiling, and can be destroyed by carbolic acid and by corrosive sublimate. The streptococcus of erysipelas is thought to be identical with the streptococcus pyogenes, but their difference in action is believed to be due to difference in virulence induced by external conditions and by the state of the tissues of the host. The coccus of erysipelas is larger than the ordinary form of streptococcus pyogenes, and infection takes place through a wound, often a very trivial one, or through a mucous membrane. The organism multiplies in the small lymph channels. The streptococcus may cause suppuration in erysipelas, mixed infection not being necessary to cause pus to form.

The *gonococcus* of Neisser is found both inside and outside of pus cells and mucous cells. The gonococci cannot be cultivated upon ordinary media, but grow best upon human-blood serum. Gonococci stain easily and are readily decolorized by Gram's method.

The *bacillus coli communis,* or the bacillus of Escherich, is invariably found in the fæces. It is believed by many observers to be the cause of appendicitis, peritonitis, and abscesses about the intestine. In cases of appendicitis we can rarely get a pure culture of Escherich's bacillus, but usually find also streptococci and staphylococci.

The *bacillus of typhoid fever* (Eberth's bacillus) is responsible for some cases of gangrene, for some of embolism and for not a few bone and joint diseases.

The *bacillus tuberculosis* (Koch's bacillus), the cause of all tubercular processes, is met with especially in dusty air which contains the dried sputum of victims of tuberculosis. This infected air is the chief means of its transmission, though it may be conveyed by the milk of tubercular cows and by the meat of tubercular animals. Wounds may open a gateway for infection.

The *bacillus tetani* (Nicolaier's bacillus), an aerobic organism, is found especially in the soil of gardens, in the dust of old buildings, in street dirt, and in the sweepings of stables. Spores develop at the ends of these bacilli. This organism is capable of producing toxins of deadly power. Its spores are hard to kill.

CHAPTER III

ASEPSIS AND ANTISEPSIS

Before the introduction of Lister's methods of treating wounds, it was considered proper, in accidental and operative wounds, to have profuse suppuration, pyemia, erysipelas, etc., and it was not remarkable, therefore, that the mortality following accidental and operative wounds was very high. Lister's method of wound treatment was largely based upon the conception that the infection of wounds occurred from contact with the air which contained spores and germs, and his method of treatment therefore, was directed chiefly to their destruction. The air can be a medium of wound infection to a certain extent, and dry air contains more spores and bacteria than moist air, but Koch demonstrated the fact that atmospheric microbes were chiefly of innocuous character, and wound infection usually could be traced to bacteria or spores being brought into direct contact with wounds, by the clothing, or by the skin of the patient, or by the hands of the surgeon, or by unclean surgical instruments and dressings. The antiseptic qualities of the blood serum and cell activities in healthy tissues, are sufficient to destroy or remove a certain number of microorganisms, and suppuration occurs only when the tissues are completely overwhelmed by the number of these organisms or when their power of resistance is lessened by injury or disease.

Sepsis. Sepsis is due to the entrance and multiplication of microbes, or to the absorption of their products in the body. Local inflammation and marked constitutional symptoms characterize sepsis.

Asepsis. Asepsis aims at thorough sterilization of the parts and of all the objects brought into contact with the wounds, and the exclusion of germs by the use of occlusive bandages and dressings.

Antisepsis is that method of wound treatment which keeps germicidal agents continuously in direct contact with the wound. Its object, therefore, is to produce asepsis. It is the duty of the surgeon to guard against the contact of microorganisms in the wound and to employ whatever means science has evolved for their destruction. He must, however, be careful to employ means of disinfection or destruction that will not have an injurious effect upon the normal tissues.

Mechanical disinfection does not apply to wounds but is employed as a preventive measure by the operator and his assistants for instruments and for the skin surrounding the wounds. Mechanical disinfection is accomplished by the use of soap and water and a friction brush; germicidal solutions of one kind or another are also employed.

In the modern aseptic operating room germicides and antiseptics do not play so important a part as they formerly did. This is largely due to the fact that heat is used wherever possible in the preparation of sutures, ligatures, dressings and instruments, and to the farther fact that in uninfected tissues no antiseptic solutions are employed. It must also be remembered that the germicidal agents possess the disadvantage of exercising a more or less destructive action on the body cells, and consequently their use is not warranted in clean wounds. We still, however, sufficiently often meet with infected wounds that render the use of these agents necessary.

Heat is the most valuable of all sterilizing agents, its only drawback being that it is not universally applicable. Wherever possible it should be employed in preference to chemical agents. It can be employed either dry or moist. Moist heat is a much more efficacious germicide than dry heat, for it destroys the organisms at a much lower temperature. Boiling water at a temperature of 212°F. will destroy nearly instantaneously all pus-producing organisms. Spores, however, require a moist heat of 284°F. kept up for at least a half-hour. A dry heat of 212°F. will not destroy pus-producing organisms under an hour and a half of treatment, and spores will live for three hours at a dry temperature of 284°F.

Although moist heat is very much quicker and more satisfactory in its action, yet it is often inconvenient to employ it in the sterilization of gowns, towels, operating suits, etc. However, sterilization by heat has been greatly facilitated by the introduction of the autoclave, by means of which a very high temperature under pressure can be obtained. This is the most satisfactory method of sterilizing dressings, towels, sheets, operating suits and aprons. A similar and less expensive method of sterilizing these articles is by the use of one of the simple steam sterilizers which are sold by all dealers. In an emergency, an ordinary bake oven can be employed as a sterilizer. It is best, however, where the temperature cannot be estimated, to boil the articles and dry them between sheets moistened with bichloride solution.

Disinfection or Sterilization. Sterilization of a wound, or of the substances coming in contact with it, may be accomplished by using the aseptic or antiseptic method; by combining these two methods we obtain the best results. The aseptic method, which employs antiseptic substances for the purpose of sterilization of objects coming in contact with the wound when their disinfection by heat is impossible, is the method perhaps most generally favored by modern surgeons.

Antiseptic Method. In the antiseptic method, the field of operation, the hands of the operator and of his assistants, and the instruments, must be treated in germicidal solution and, in addition, the wound should be frequently irrigated during the operation with a solution that has germicidal properties.

Recent investigations show that many germicidal substances have not the power that was formerly attributed to them. Furthermore, substances which are really active germicides very often produce a marked toxic effect upon the patient and produce a very decided irritation of the skin with which they come in contact.

Aseptic Methods. The aseptic method for the treatment of wounds admits of the use of germicidal solutions and heat upon the field of operation, upon the hands of the operator and of his assistants, and upon the instruments employed. After this has been accomplished, placing absolute dependence upon this sterilization, no germicidal or antiseptic substances are brought into contact with the wound, sterilized salt solution or plain sterilized water being used, if necessary, to flush the wound, the dressings employed having been sterilized by dry heat or moist heat.

Sterilization of the Hands. Experimental investigation has shown that the failure of the surgeon's efforts to render his hands absolutely aseptic, has been the productive cause of infection in many wounds.

The hands and finger nails may be best sterilized by first rubbing them with spirits of turpentine; then scrubbing them with soap and water; and then using a sterilized nail brush freely. The scrubbing should be done for several minutes. The hands should then be rinsed to remove the soap, and then soaked for about ten minutes in a solution of bichloride, strength, 1 to 2500. If turpentine has not been used before washing with the soap, strong alcohol or ether should be well rubbed over the hands before they are immersed in the bichloride solution. Perhaps the best way of rendering the hands sterile is to scrub them with green soap and water, then mix a tablespoonful of commercial chloride of lime and half a tablespoonful of carbonate of soda with enough water to make a paste. When this has assumed a thick creamy consistency, it should be rubbed into the hands until the grains of lime disappear and the skin feels cool; then rinse the hands in sterile water.

Sterilization of Instruments. Instruments may be sterilized by boiling them for fifteen minutes in water in which a tablespoonful of washing soda has been added for each quart. This prevents rusting of the instruments and also makes the water a better solvent for any fatty matter which may be upon the instruments, thus increasing the sterilizing effect of the heat.

Sterilization of the Feet. As most patients do not apply water as freely or as frequently to the feet as to other portions of the body, there is usually present an excessive amount of thickened epidermis, which is very difficult to render sterile. For operations in chiropody the feet should be thoroughly moistened with soap and water, scrubbed vigorously with a brush, then soaked in a solution of bichloride of mercury of 1 to 1000 strength, and then wrapped up in a towel soaked in the same solution while waiting for the operator.

AGENTS EMPLOYED TO SECURE ASEPSIS

Bichloride of Mercury is used for the disinfection of the hands and skin and for the irrigation of wounds. Biniodid of mercury is extensively employed and in the same strengths as the bichloride. It is, however, a more powerful germicide, while being less irritative, and neither forms a mercuric albuminate nor tarnishes metal instruments.

Carbolic Acid. This acid is derived from coal tar, and although known as early as 1834 as the first antiseptic recommended and used by Lister, is not so popular since the discovery that bichloride of mercury possesses more germicidal action.

Gangrene of the skin and subjacent tissues has often been traced to the long continued use of dilute solutions of carbolic acid or of ointments containing small quantities of the drug. Gangrene of the fingers and toes is by no means infrequent as a consequence of its use. Another condition frequently seen is the systemic poisoning through absorption. One of the first symptoms noticed from such absorption is irritation of the urinary tract and carboluria. This poisoning is more apt to take place when the weaker solutions are used than when the pure acid is used, as the destruction produced by the pure acid prevents its absorption.

The effect of carbolic acid upon the urine (See Chapter II, "*Carbolic Acid*") is to cause it to become smoky a short time after it is voided. The urine shows a complete absence or diminution of the sulphates, and albumin is generally present. When these symptoms present themselves, the use of carbolic acid should be withdrawn, and the administration of sulphate of soda and atropin begun. If the condition has existed for any length of time and the patient is weak and exhausted, stimulants are indicated.

Lysol is a saponified phenol, and possesses some germicidal power. It is used in strengths of 1 to 3 per cent. solutions.

Creolin is mildly germicidal and is used a great deal in from 2 to 4 per cent. solutions.

Both lysol and creolin act very much like carbolic acid, but neither possess its irritating qualities.

Formaldehyde Gas is an active germicide and very valuable as a disinfectant. It is used in the shape of formalin which is a 4 per cent. solution of the gas in water. This agent is very irritating to the normal tissues in the stronger solution, but a 2 per cent. solution of formalin may be used for the sterilization of the hands, instruments, etc.

The formaldehyde fumes are employed for the disinfection of clothing, rooms, bedding, and also for the sterilization of catheters. The fumes of the gas are very irritating to the mucous membrane and when this agent is used for the disinfection of rooms, every crevice and crack must be tightly sealed to prevent the escape of the gas.

Iodoform. The action of iodoform is not due directly to its ability to destroy germs but to its undergoing decomposition in the presence of moisture, liberating iodin and thus rendering inert ptomains that have resulted from the growth.

Iodoform Powder is rapidly absorbed by the skin and fatal cases of iodoform poisoning have occurred from treating burns with it. Iodoform is also used in ointment form and in suppositories. As it is insoluble in water it is commonly used in a 10 per cent. emulsion. The gauze is also greatly used.

The symptoms of iodoform poisoning are: delirium; odor of iodoform on the breath; presence of iodoform in the urine; eruption over the skin, and finally, coma. Iodoform is also capable of producing a localized dermatitis, with great irritation, and must therefore be used with care on all delicate skins.

Aristol, a substitute for iodoform, is a compound of iodin and thymol, producing no toxic effects and having no disagreeable odor; it does not, however, possess the germicidal qualities of iodoform. Nosophen, iodol, and airol are among the more recent substitutes.

Iodin. This drug no doubt possesses more germicidal properties than was at one time supposed. It is probably the most powerful antipyogenic known. The 7 per cent. tincture is the one most frequently used.

Acetate of Aluminum, or more properly, aluminium, is prepared by adding five parts of sugar of lead to a solution of five parts of alum in 500 parts of distilled water. Burow's solution, see page 35, is chiefly employed as a wet dressing.

Chloride of Zinc in a solution of 15 to 30 grains to the ounce, has marked antiseptic properties, but it blanches the tissues when applied to infected wounds.

Sulphocarbonate of Zinc is less irritating than the chloride of zinc and is of the same value as a germicide.

Peroxide of Hydrogen when used as a 15 volume mixture or diluted, seems to have a direct action upon pus generation by destroying microorganisms of the pus. It is frequently employed for sterilizing abscess cavities, and for hastening the separation of necrotic tissue.

This agent has also a marked hemostatic power and is used to some extent on this account in nose and throat work. Its hemostatic power is also observed in bone cavities. Care should be taken never to use it unless there is a free exit, as it increases rapidly in volume after coming in contact with dead tissue or pus, and serious accidents have happened from its improper use; for instance, if it is injected into an abdominal sinus where free escape is not provided for, the distention will result in ruptures of the sinus and infiltration of the surrounding tissues; possibly of the peritoneal cavity. The distention produced by it is also quite painful and therefore only a small quantity, or a much diluted solution should be introduced into cavities.

Boric Acid is not very actively antiseptic, but even in a saturated solution it is not irritating. Where bichloride or carbolic dressings have produced irritation of the skin, or burns, a boric acid ointment is a very satisfactory substitute.

Salicylic Acid is an antiseptic of value. It is generally used in the form of an ointment. It is but slightly soluble in water.

Potassium Permanganate by its rapid liberation of oxygen, acts as an antiseptic of proven merit for the disinfection of foul wounds and ulcers. It is also used satisfactorily for disinfecting the hands in preparation for operations, in the form of a 5 per cent. solution, any stain being removed later by a saturated solution of oxalic acid.

Alcohol possesses marked antiseptic properties and is one of the best agents for the sterilization of the hands of the surgeon, and for the skin of the patient. A 60 or 75 per cent. solution of alcohol is much more efficacious as a skin disinfectant than a 95 per cent. solution. This is because the purer alcohol is much less penetrating than the dilute. It is also used when diluted with water, one part to four, as a dressing for granulating wounds. It is efficacious in limiting the action of carbolic acid, when this agent has been applied in full strength.

It is a useful agent in which to store certain materials such as ligatures, sutures, etc.

Silver Nitrate possesses undoubted antiseptic properties, and solutions of varying strengths are decidedly antiseptic. These solutions are from 5 grains to the ounce, to 60 grains to the ounce.

The solid stick of nitrate of silver is used for destroying exuberant granulations. Among the different silver preparations on the market, protargol and argyrol are the best known. Both of these are extensively used in the treatment of inflammations of the mucous membranes.

The unguentum of Crede, is an ointment of silver which is used in cases of septic infection and also in localized inflammations. From 15 to 45 grains of silver can, in this form be rubbed into the skin. It is absorbed and undoubtedly exercises an antiseptic influence on the infecting microorganisms.

Saline Solution, or normal, or isotonic salt solution, as it is called because of its close approximation to the blood serum, consists of a solution of 7 per cent. of sodium chloride in plain sterilized water. Roughly speaking and for ordinary purposes, this solution can be made by adding an even teaspoonful of ordinary table salt to one pint of boiled water and then reboiling the mixture.

It can be stored for a limited time in sterile glass jars, which are sealed with sterile cotton. The jars can be heated to whatever temperature is required for use. This solution is the one which is generally used for irrigating wounds and cavities; it is non-irritating and possesses no antiseptic quality. When a moist dressing is desired there is no solution comparable to it, largely because of its non-irritating quality. It has at times a slight irritating effect upon the kidneys and when large quantities of it are used it is better to dilute it.

Pure Oxygen and Ozone have been used, and the latter is more effectual. It has been found that oxygen but slightly retards the growth of bacteria, but both ozone and oxygen produce a hyperemia, and retard the growth, especially of anaerobic organisms. Pure oxygen in the abdominal cavity produces a marked hyperemia and a leukocytosis. Ozone has been put to some practical use in this country but the results have not been sufficiently studied.

Sunlight has a marked retarding effect on some bacteria and actually destroys them. The anthrax spore is said to be killed very promptly by exposure to strong sunlight and it is claimed that the tubercule bacillus is slowly destroyed by it.

Electricity and the X-rays also produce a marked retarding effect on the propagation of certain microorganisms.

CHAPTER IV

INFLAMMATION

Definition. Inflammation may be defined as the local reaction against injurious influences. An aseptic wound heals without any of the clinical signs of inflammation and without reaction. It is only by a study of the minute changes about such a wound that the resemblance, between the processes of wound repair and those of slight inflammation, become evident.

Etiology. The cause of inflammation is any injury to the tissues by mechanical, thermal, or chemical means; by the effect of electricity, or by the growth of bacteria.

Pathology. Inflammation occurs through changes in the circulation.

When one of the causes mentioned above acts upon the tissues, the first alteration seen is an increasing blood supply to the part, the arterial circulation being increased both by the greater rapidity and force of the current through the vessels, and by the dilatation of all the small branches and capillaries.

When the inflammation grows more intense, the circulation in the capillaries becomes slower and the corpuscles collect, until they clog the vessels. The normal current of blood in small vessels, as seen under the microscope, shows a thick central stream of corpuscles with a transparent border of lymph (containing only a few white corpuscles) between it and the vessel wall.

As the stream diminishes in rapidity, the number of white cells in the clear space increases, the blood plaques appear also, and finally, when the current is reduced to stagnation, the clear space disappears, being filled entirely with cells, chiefly leucocytes, although red cells find their way into it.

This tendency of the white cells to separate from the others, even when the current is rapid, is partly due to their viscosity and power of ameboid movement, but in the main is a purely mechanical effect of the slower current.

It has been proven that when particles of different density are suspended in a liquid which is circulating through a system of narrow tubes with a very rapid current, there is a clear space next to the wall of the tube where the friction necessarily reduces the speed of the fluid which is free from particles, and, as the current is slowed down, some of the particles of least density, begin to appear in this clear space, their number increasing as the current becomes slower, until even the heavy particles also collect here when it is very slow.

It is known that among the cellular elements of the blood, the leucocytes have the least specific gravity or density, and the blood plaques rank next, while the red blood disks are the heaviest, and these bodies appear in the clear serum near the vessel wall in that order, according to the law just cited. The slow current is associated with an increased intravascular blood pressure, which, in part, is the cause of the phenomena of exudation, emigration and diapedesis.

Exudation. Serum of the blood passes out of the vessels, and collects in the lymphatic spaces in the cellular tissue, and elsewhere, and also exudes from the surface of the mucous membranes or forms vesicles or blisters in the skin by detaching the superficial epithelial layers. Complete stasis, or stoppage of the circulation is seen only when the inflammation is exceedingly intense, and would cause the death of the part if continued long.

Usually the current merely becomes slower than normal. This retarded circulation is followed by the phenomena of emigration.

Emigration. Emigration of the white blood corpuscles consists in the passage of the cells directly through the vessel walls. It is most frequently seen in the capillaries, although it also takes place in the small veins. The white corpuscles, or leucocytes, have the property of ameboid movement, stretching out at will in any direction, long, narrow processes of their protoplasm, called pseudopodia, which may be attached to any object, and having secured such an anchorage, the rest of the protoplasmic body is drawn towards it.

In this way, the leucocytes are able to pass through the interstices between cells, or along narrow channels in the tissues. When the blood current becomes sufficiently slow to enable them to cling to the walls of the vessels, it is then that ameboid movement begins.

Sometimes the cells loose their hold and are swept on again, but in other cases a minute bud of protoplasm will appear on the other side of the wall of the vessel, opposite to the spot where the leucocyte is clinging, and as this grows larger, a narrow neck of protoplasm can be traced through the wall directly to the leucocyte, and presently the mass of the leucocyte becomes proportionately smaller as the external bud of protoplasm grows larger. The conditions are gradually reversed, the nuclei of the cells appear outside and only a small mass of protoplasm remains within the vessel until finally the entire leucocyte is in the tissue outside of the vessel and is free to wander in any direction.

The mechanical part of this process is not yet understood. It is claimed by some that small openings exist in the walls of the vessels, between the endothelial cells which line them, to which is given the name of *stomata*. These openings ordinarily are invisible, but they are said to enlarge under the effect of the dilation of the vessels, and of the alterations in their walls, produced by the inflammatory reaction, and that the leucocytes escape through those openings.

There can be no doubt that the emigration is due to the ameboid motion of the cell, and the discovery of the phenomenon, to which is given the name chemotaxis, affords a sufficient explanation.

This is the influence possessed by certain substances to attract or repulse ameboid cells. In some cases this attraction appears purely to be mechanical, but it is probably a chemical effect of some kind in most, if not in all, instances.

The process of inflammation produces some chemical compound which similarly causes the cells to leave the vessels, and when there is any inflammatory action in their neighborhood, to find their way by the shortest route to the seat of the inflammation.

The leucocytes direct their course through the tissues to the chief points of inflammation by reason of chemotaxis, and surround the dead tissues, or any point of bacterial growth, or any foreign body which may be the cause.

The wandering leucocytes form the pus cells, and if they are very numerous, they constitute a purulent or suppurative inflammation.

The wandering cells, however, are almost entirely made up of leucocytes, of which three forms are known, varying in size and in the size and number of their nuclei. The leucocytes surround any foreign body, and if the particles are small enough, they incorporate them within themselves, in fact, they may be said to swallow them. This taking up of particles by the wandering cells is called *phagocytosis*.

Diapedesis. When the circulation becomes very low and the pressure very high, there is a tendency of the red corpuscles to leave the vessel.

This is a purely passive process, and is observed only when the changes in the vessel wall are extreme. Both varieties of these cells die and are destroyed in the exudate, the former furnishing the fibrin which is so abundant in some forms of inflammation. This escape of red corpuscles is known as *diapedesis*, and is sometimes so extensive as to amount to capillary hemorrhage.

Symptoms. From antiquity the local symptoms of inflammations have been enumerated, as heat, redness, pain and swelling and to these has been added, impaired function.

The *redness* is due to congestion. The *pain* is due to the pressure exerted on the sensory nerves by the surrounding swelling, as is well shown by the intensification of the distress, as every beat of the heart forces more blood into the space already filled. In some cases, however, it may be caused by the direct action of the inflammatory agent upon the nerves. The *heat* is caused by the increased supply of warm arterial blood, for it has been abundantly proven that the temperature never rises above the heat of the blood, although naturally in a patient with fever, it will be above the normal temperature of that fluid. The *swelling* is due to the dilated vessels, and to the escape of serum and blood cells from the vessels into the tissues. The *impaired function* is chiefly caused by the pain which is often increased by any attempt to use the part, and by the swelling which prevents free movement, though the loss of function may also be dependent upon the direct action of inflammation upon the nerves.

The constitutional symptoms of inflammation are an elevation of temperature with or without a chill. There are also other disturbances, such as nausea, vomiting, diarrhea, sweating and polyuria. These are due to efforts on the part of the general economy to eliminate toxic substances.

The inflammatory products may poison the system in two ways: (1) by the diffusion of their chemical substances, (toxins and ptomains), or (2) by the passage of bacteria themselves into the blood.

Termination. Inflammation may result in resolution, suppuration, necrosis or sloughing, or in the establishment of a chronic state.

Resolution. Resolution is the termination of an inflammation by the gradual cessation of all the changes which have occurred. The pain subsides, the circulation becomes more normal, and the exudate is absorbed, or makes its way to the free surface of the body, where drainage occurs either spontaneously or by incision.

If there has been any loss of substance caused by the inflammation, it is restored by processes exactly similar in character to those in the repair of wounds.

Suppuration. Pus consists of a serum containing little or no fibrin and large numbers of leucocytes. There are also many cells, either dead or dying, which represent the waste thrown off from the tissues as a result of the inflammatory reaction. A purulent inflammation or suppurative inflammation, is one in which there is pus formation.

When suppuration occurs, the pus may make its way to a free surface, such as a mucous membrane, or may form an abscess, or may cause sloughing of the skin over the seat of inflammation, and so escape from the cellular spaces in the tissues.

Pus may be thrown off by a mucous membrane, without any actual breach of continuity. Diffuse infiltration of the tissues is the most dangerous form of suppuration.

In this variety of inflammation the exudate is brought into contact with the greatest possible extent of absorbent vessels, for as a surface of a sponge is greater than that of a bag, which would contain it, so the surface of these intercellular spaces is much greater than that of an abscess cavity filled by the same amount of pus. In this form the bands of cellular tissue, lying between and forming the boundaries of these spaces, remain intact, and the exudate is either absorbed into the circulation, or seeks escape through many punctate openings in the skin.

The entire skin of the part is frequently detached from the fascia by the sloughing of the subcutaneous tissues, before it gives way, and even when it finally yields to the necrotic process, the openings formed will be altogether too small in proportion to the extent of the disease beneath, so that healing is still further delayed.

Sloughing. Inflammation may be accompanied by sloughing or death of tissues. Gangrene, mortification or necrosis is a death of the tissue from any cause. The part which has died is designated as a *slough.*

When inflammation has subsided, granulation tissue forms on the living tissue, exerting pressure upon the slough, thus hastening its absorption or separation.

Chronic Inflammation. An interruption at some stage of resolution or suppuration and the continuance of mild symptoms constitutes a chronic state.

By chronic inflammation, we understand a long continuance of some or all of the changes seen in acute inflammation, but less in intensity, and an abnormal tendency to the production of new tissue.

Treatment. The general indications to be observed in the treatment of inflammation are: (1) to combat the congestion of the parts; (2) to relieve tension; (3) to give free issue to the products of inflammation; (4) to produce early separation of sloughs.

Very hot or very cold applications exert a beneficial and soothing effect upon inflamed areas.

Cold has the tendency to reduce tension by constricting the blood vessels thus diminishing the amount of blood supplied. In an infected area the reproduction and development of bacteria are checked, and suppuration is frequently aborted.

Heat has the effect of dilating the blood vessels and hastens repair in bruised, strained, or torn tissues. This is a variety of hyperemia treatment which is especially useful in the absence of bacteria. In infected areas the growth of bacteria, and increased pus formation, would be encouraged and heat is contraindicated.

We are yet without an antiseptic material which can be used in sufficient strength to affect the growth of germs and yet not injure the patient. Injury of the part treated, and absorption into the circulation are both to be avoided. The application of dressings, wet with corrosive sublimate, or other chemical solutions to the unbroken skin over inflamed areas, is a fallacy. Any benefit which has been observed to follow their use, has undoubtedly been due to the effect of the moisture and warmth or cold, according to the temperature of the dressing, thus obtained, while local sloughing and general constitutional poisoning are a common result of such applications. A light gauze dressing, applied cold, and kept constantly wet with any evaporating solution, will greatly relieve the congestion and so assist the inflamed tissues in their contest with any irritating materials.

A thick wet dressing made with a hot solution, and well protected against evaporation so that it will retain its heat, will produce the same effect as a poultice, although less powerful. When there are discharging wounds or raw surfaces, unprotected wet gauze should be employed, for poultices are then inadmissible, and the weak antiseptic solution will inactivate and wash away bacteria.

Astringent solutions have an excellent effect upon inflammatory processes and the most generally useful of these is the 50 per cent. solution of acetate of aluminium.

The following is a modified Burow's solution:

Alum	24	gms., or	6	drachms
Lead acetate	38	" "	9½	"

Water 1000 " " 2 pints

Filter after mixture has been allowed to stand for 24 hours.

Ointments are employed by many in the treatment of small areas of inflammation; they are useful, though not as efficient as hot or cold wet dressings. Over the unbroken skin, they can only act like a poultice and should not be employed where infection exists. On clean wounds they are unnecessary, but upon ulcers or wounds which show no tendency to heal, such ointments as Peruvian balsam, 5 per cent., or scarlet red, 4 per cent., are extremely valuable.

THE PROCESS OF REPAIR

Regeneration of Tissues. The reparative powers of the tissues of the human body are considerable, although not comparable with those of the lower animals, in the lowest orders of which the reproduction of an entire limb, or even one-half of the body, may take place. In order to understand the regeneration of tissue, we must first consider briefly the life history of the cells.

A cell consists of a mass of protoplasm, generally enclosed in a cell membrane, and containing a nucleus and nucleolus. The nucleus represents the most vital part of the cell protoplasm, and has a more granular appearance than the latter. The nucleolus is a minute solid spot in a nucleus, appearing to be more highly refractive.

Cell Division. When the cell is quiescent, the protoplasm appears evenly granular, but when it is stirred to active life, slender twining threads can be traced in the nucleus, perhaps consisting of one long thread twisted upon itself.

On account of their readiness to take up dyes used in staining, these threads are called *chromatine threads.*

When the cells are about to divide, the chromatine threads are seen to arrange themselves in a line across the center, called the *equator* of the nucleus, forming a rosette or star shape, known as the *mother star*. Some large granules then appear in the nucleus at points on either side of this line, which are known as the *poles* of the nucleus. The loops of the thread are directed towards the poles. Gradually these threads become arranged in radiating lines, converging at the poles, and then break away from their former connections with the equator, forming a*daughter star* at each pole, a clear space appearing at the equator. A constriction next appears in the now clear equator, and the nucleus divides into two distinct nuclei. Simultaneously with this division, or immediately following it, the protoplasm of the cell body divides in the same place, and thus two complete cells are produced. The chromatine threads lose their rosette arrangement, and gradually become imperceptible as the new cell returns to the quiescent state. This process of cell division is known as *karyokinesis* or *aryomitosis*.

In simple cells like the leucocytes, reproduction may take place by simple fission, thus: a constriction appears in the nucleus and in the body of the cell in the same line, and the two divide without any visible protoplasmic changes. Such a simple mode of division does not occur in the more highly specialized cells of various tissues. If the karyokinetic action be not very vigorous, the nucleus may divide, but the cell body remains intact, producing the cell with two or more nuclei so commonly observed. Every cell reproduces its kind, spindle cells producing connective tissue; epithelial cells epithelium; and bone cells producing bone.

Repair of Wounds and Healing by Apposition. When a wound occurs, the cut edges immediately retract on account of the elasticity of the tissues, and the gap fills with blood and serum. If no bacterial or chemical irritant is introduced, there are no true inflammatory changes. The divided blood vessels are soon plugged with coagulated blood, which extends into the cut vessels to the nearest branch. The capillaries around the seat of injury dilate slightly, the fixed cells of the tissues become active, dividing by karyokinesis as already described. The endothelial cells lining the divided blood vessels multiply and take an active part in the process. In spite of the congestion and the new cells produced, the reaction is much less than that of inflammation. The new cells invade the blood clot, consuming it and also any foreign matter, or any tissue which may have been killed by the injury. From the loops of the occluded capillaries, at the sides of the wound, spring buds of endothelial cells, becoming thicker and then hollow as they extend, blood cells forming in them and blood entering them also from behind. These advancing endothelial tubes join with those on the opposite side of the wound, and thus the new forming tissues are supplied with blood vessels.

It is said that new vessels are also formed by the pre-existing lymph-spaces and by independent cells. Meantime the connective tissue cells have been forming fibres across the clot and epithelial cells over its surface, if skin or mucous membrane be involved in the injury. The new vessels disappear, and the new connective tissue forms the scar. This is the process of primary union in a wound in which there is not a marked cavity or a loss of tissue on any of the exposed surfaces of the body, and no matter how closely the edges of such a wound may lie in contact, it can heal by no other method. Even the closest apposition of the sides of a wound cannot prevent the interposition of a thin layer of clot and the partial death and absorption of a very thin layer on its surfaces. This is also known as primary union.

Healing by Granulation. When a wide gap has been produced by retraction or by actual loss of tissue, healing takes place by granulation, as it is called, a process which differs from that just described merely in the fact that more tissue must be reproduced. The outpouring of blood and serum, occlusion of the vessels, congestion, multiplication of fixed cells, emigration of leucocytes, and production of vascular loops and buds, goes on as before. As the formative changes advance, small, round elevations of a rosy color appear on the new surface, making it look like velvet. These rounded elevations of the healing surface are called granulations.

They advance steadily on all sides, filling the gaping wound until the level of the original surface is reached, the new tissue organizing behind them, and contracting as it organizes, so that the space to be filled is daily made smaller by this contraction as well as by the production of new tissue. As the surface is reached, the epithelial cells on the edges of the granulating area slowly spread over it, the granulations generally projecting above the adjoining surface and the epithelium growing over them as they contract again to their proper level. The advancing line of epidermis is visible as a pink line, gradually whitening with time.

CHAPTER V

WOUNDS AND CONTUSIONS

A wound is a solution of continuity or division of the soft tissues produced by cutting, tearing, or compressing force. The classification of wounds according to their causation or nature is as follows:

Incised—when resulting from a sharped-edged instrument.

Lacerated—when tissues are extensively torn or separated.

Contused—when resulting from a more diffused force, tearing and bruising the tissues.

Punctured—when produced by a narrow instrument that causes a wound deeper than its external surface is broad.

Poisoned—when some poisonous substance enters the wound and causes local infection or constitutional disturbance.

Gunshot—when the injury results from firearms or powder explosion.

An Incised Wound is an injury which is produced by some sharp instrument such as a knife, pieces of glass or metal, which divides the tissues cleanly, producing no bruising or tearing. The pain is usually sharp and burning, varying with the nature of the instrument with which the injury has been inflicted. Hemorrhage is usually free.

Lacerated Wounds. These usually result from machinery accidents or from heavy bodies passing over the parts and are apt to contain a considerable quantity of foreign matter ground into the tissues.

Contused Wounds. A contused wound is one in which the edges and surrounding tissues are bruised or crushed. External bleeding as a rule is not excessive, although there is a great likelihood of extensive subcutaneous hemorrhage. Sloughing and gangrene may occur.

Punctured Wounds. The character of a punctured wound depends upon the object producing it. If made by sharp instruments, such as knives, swords, daggers, bayonets, or needles, their nature is similar to incised wounds.

Unless organs of importance have been wounded, or unless active septic material has been carried into the wound, healing promptly follows after the withdrawal of the instrument which has caused the wound. These wounds are usually deep when affecting the dorsal aspect of the foot, being commonly caused by a falling instrument or tool. In the plantar region they are of every degree of severity, from the most minute puncture to perforation running between interosseus spaces and passing through the dorsal skin. The most frequent punctures are those caused by stepping upon needles, pins and tacks. These wounds are, commonly, of no importance unless the foreign body is broken off or entirely penetrates the foot.

If the patient is seen a very short time after this has occurred, the surgeon may operate with some confidence of finding the offending substance, but even here, if possible, it is an advantage to obtain an X-ray picture, while in those cases in which a needle has long been buried in the tissues, this is quite indispensable. It is well to remember that in these cases the patients' impressions us to the location of the needles are most unreliable.

After a radiograph has been obtained, it is most important, if anatomically possible, to make the incision at right angles to the shaft of the needle. At least two pictures should be taken in order, if possible, to obtain some idea of the depth at which the needle lies. Even with all these helps, the procedure, simple though it may at first appear, oftens turns out to be one of great difficulty, necessitating a very extensive operation.

Incised Wounds of the Foot. Incised wounds of the dorsal surface are very frequently quite deep and often implicate the tendons, bones and articulations, as they are most frequently inflicted by the fall of some heavy tool upon the part, or by the inaccurate blow of an axe. Wounds of slight importance need but the usual thorough cleansing out, with or without suturing of the skin, according to the extent of the incision.

If one or more of the tendons have been severed, the ends should be approximated by catgut sutures. If extensor tendons are cut in the neighborhood of the metatarsophalangeal joints, it is often necessary, owing to considerable retraction of the distal end, to incise the skin down as far as is needed, in order to secure the retracted end and suture it. Failure to adopt this procedure permits a dropping of the toe, converting it often into a regular hammertoe. When the tendon is properly sutured, the toe must be placed for some days in a condition of over extension, most easily secured by a bandage passed under it, acting like a stirrup, the ends being fastened by several turns above the ankle.

Incisions, implicating joints, are carefully cleansed by flushing the joint with copious quantities of saline solution, and closing the wound with very few stitches. Such injuries should be examined daily and any sign of sepsis must be considered as an indication for immediate removal of the stitches, followed by active antiseptic wet dressings.

Cuts of the plantar surface are not often very extensive. They are most frequently incurred in stepping upon some sharp instrument or walking upon glass, especially while bathing.

Contusions. A contusion or bruise is a subcutaneous laceration, the skin above it being uninjured, as in the abdomen; or being damaged without a surface breach, as in a part overlying bone, and blood being effused. If a large vessel is damaged, hemorrhage is extensive.

An *ecchymosis* (black and blue area) is diffuse subcutaneous hemorrhage.

A *hematoma* is a blood tumor or a circumscribed hemorrhage in the tissues.

In a diffuse hemorrhage the coagulation of fibrin induces induration, the serum and leukocytes are absorbed, the red blood cells disintegrate, and the coloring matter is widely diffused by the tissue fluids, and hemoglobin is changed into hematoidin which crystallizes. In union with these chemical changes, color changes ensue, the part being at first red and then becoming purple, black, green, lemon and citron. The stain following a contusion is most marked in the most dependent area.

A hematoma acts as an irritant, inflammation ensues around it and it is encapsuled by embryonic tissue, which, by organizing into fibrous tissue, forms a blood cyst and gradually absorbs the fluid blood, the cysts contents becoming thicker and thicker. A fibrous scar may remain, and a blood clot, with very much indurated surrounding tissue, giving a hard edge, is noticed after bruises of the periosteum. If serum is not absorbed, hematoidin forms and the fluid becomes clear. A hematoma may suppurate, an abscess forming, but this rarely happens except in drunkards, although it occasionally occurs in persons who do not use alcohol.

Symptoms. The symptoms are tenderness, swelling, pain, and numbness. The pain may be severe, but rarely persists beyond the first twenty-four hours. Discoloration appears quickly in superficial contusions, but only after days, in deeper ones. Shock and loss of function are present only after severe contusions. The swelling is first due to blood and is soon added to by inflammatory exudation.

Terminations of Contusions. Slight contusions terminate promptly by resolution; the more severe may terminate in gangrene, inflammation, abscess, fibroid thickening, hypertrophy of the tissues involved, (as in the case of bone), chronic inflammations, and even malignant growths, particularly sarcomata.

Prognosis. The prognosis of contusions is a matter of every day importance, and it is sometimes extremely difficult to prognosticate accurately. The determining forces are principally the nature and violence of the contusing force, the tissues and organs involved, and the general condition of the patient. Even the injury of the tissues that may be easily inspected, such as the skin, may be much more severe than is apparent. In tissues of low vitality, such as synovial membrane, cartilage and ligaments of a joint, repair is proportionately delayed, whereas in highly vascular tissue it is more rapid. Contusions of tissues that cannot be given physiologic rest, such as the thoracic wall, and the respiratory muscles, respond less promptly to treatment.

The general condition of the patient is an important factor in the prognosis, the most favorable being vigorous adult life without organic disease. Among the unfavorable general states are, the extremities of life, the very anemic and the plethoric, the tuberculous, the syphilitic, the diabetic, and like diatheses, while in the rheumatic and the gouty, the slightest injury may be most persistent. The starved, the overfed, the over-worked, the fatigued, the alcoholic, and those exposed to extremes of heat and cold, are unfavorably affected.

Treatment. Slight bruises, favorably located, require no treatment. The arrest of hemorrhage, thereby diminishing the swelling, pain, and discoloration, is important. If the hemorrhage be from small vessels, elevation, rest, and the application of ice are sufficient. Frequently the application of pressure is indicated. Hemorrhage in deeper parts, such as that occurring under the fascia of the thigh, is sometimes best controlled by adhesive strapping. If the vessels are large and the hemorrhage is rapid, it is sometimes necessary to make a free incision and apply a ligature. Evaporating lotions or elastic pressure by bandaging over absorbent cotton, may assist. If the hemorrhage be in a joint causing immediate swelling, painful from distension, prompt aspiration will give relief. This should only be resorted to under the strictest aseptic precautions, as the conditions are favorable for microbic growth. If the soft parts are so severely contused as to jeopardize the nutrition, both bandaging and ice should be withheld, and in some instances even warm applications are advised. After the acute symptoms have passed, judicious massage may be most helpful in securing early resolution. Restoration of the vasomotor tone when impaired or lost may be greatly facilitated by douching with cold and hot water alternately followed by massage. During the acute stages, physiologic rest is important; the restoration of functional use in severe cases must be tentative, guided by the response of the tissue in the form of increased pain or swelling. These phenomena should be avoided if possible. If hematomata be not absorbed they should be aspirated and pressure applied before structural changes take place, such as the formation of a membrane. If the latter occurs and sufficient time has elapsed for the formation of definite new tissue, aspiration may be followed by the obliteration of the sac. Sometimes hematomata become so thoroughly and firmly organized and gradually increase in size, that it is extremely difficult to differentiate them from new growths. If pain and tenderness persist for a long time, particularly, if there be a predisposition to tuberculosis, especial care is necessary.

Treatment of Wounds in General. Arrest hemorrhage, bring about reaction, remove foreign bodies, asepticize, drain, coaptate the edges and dress, secure rest to the part and combat inflammation.

Constitutionally, allay pain, secure sleep, keep up the nutrition and treat inflammatory conditions.

Arrest of Hemorrhage. To arrest hemorrhage the bleeding point must be controlled by digital pressure until ready to be grasped with forceps; it is then caught up and tied with catgut or aseptic silk. Slight hemorrhage stops spontaneously on exposure to air, and moderate hemorrhage ceases after the vessels are clamped for a time; an injured vessel of some size must be ligated, even if it has ceased to bleed.

Capillary bleeding is checked by hot water compresses. In bringing about reaction from shock, raise the feet and lower the head, unless this position causes cyanosis. At least place the head flat and the body recumbent. Apply hot water bottles and hot blankets and give hypodermic injections of ether, brandy, strychnine, digitalis or atropin, or inhalations of amyl nitrate. Strychnine can be used in large doses, one-thirtieth of a grain may be given every ten or fifteen minutes, until three doses have been taken. If the skin is very moist, atropin is indicated, alone or combined with strychnine. Hot coffee, or other hot fluids, should be given by the mouth and rectum, and mustard should be placed over the heart, spine and shins. The use of hot and stimulating rectal enemata is very important. The rectum may absorb when the stomach refuses to do so. Enemata of hot normal saline solution are very beneficial.

Enteroclysis. The tube is carried into the sigmoid flexure and the injection is introduced so as to distend the colon. At times it may be necessary to give an intravenous injection of saline solution in order to overcome the shock. In order to prevent the suppression of urine, it may be necessary to administer diuretics.

Removal of Foreign Bodies. Remove with forceps, all foreign bodies visible to the eye: splinters, bits of glass, portions of clothing, dirt, etc.

In a lacerated or contused wound, portions of tissue injured beyond repair should be regarded as foreign bodies and should be removed with scissors.

Cleaning the Wound. If the surface is hairy it must be shaved before the scrubbing. An accidental wound is infected and must be well washed out with an antiseptic solution. A clean wound, made by the surgeon, need not be irrigated, in fact, irrigation with an antiseptic fluid leads to necrosis of tissues, causes a profuse flow of serum and necessitates drainage. If clots have gathered in a wound, they must be removed, as their presence will prevent accurate coaptation of the edges. In an infected wound, they are washed out with a stream of corrosive sublimate solution. In a clean wound, they are washed out with hot salt solution. If dirt is ground into a wound, as is often seen in crushes, pour sweet oil into the wound, rub it into the tissues, and scrub the wound with ethereal soap. The oil entangles the dirt and the soap and water remove both dirt and oil. After the rough cleansing, irrigate with corrosive sublimate solution. In some cases, especially in bone injuries, it is necessary to scrape the wound with a curet.

A granulating wound is treated the same as an ulcer and the treatment is discussed under that chapter.

Drainage, Closure and Dressing. Superficial wounds require no special drain, as some exudate will find exit between the stitches and the rest will be absorbed. A large or deep wound requires free drainage for at least twenty-four hours by means of a tube, strands of horse hair, silk, catgut or gauze. An infected wound must invariably be drained. Good drainage largely compensates for imperfect antisepsis. If capillary drains be employed, apply a moist dressing. Divided nerves and tendons must be sutured. Close the edges with silk sutures or silkworm gut if the wound is deep and tension inevitable. Catgut is used for superficial wounds and for those where tension is slight. The interrupted suture is, as a rule, the best. If the wound is infected, dress with antiseptic gauze; with aseptic or antiseptic gauze if it is not infected. A dry dressing absorbs wound fluids quickly and is less likely to become infected. Change the dressings in twenty-four hours or sooner if they become soaked with the discharge. After this, in an aseptic wound the dressing need not be changed for days. If pus forms, open the wound at once.

Rest and Constitutional Treatment. In planning the treatment of wounds the most careful consideration for securing physiologic rest should be had. If at or near a joint, the parts both above and below should be immobilized. In whatever part of the body, physiologic rest should be secured as nearly as possible. If the wound be of the leg or foot, the patient should be in the recumbent position, with the part elevated and a splint applied. The factor of rest, next to that of cleansing and dressing, is most important. Physiologic rest means not only less pain, less reaction, but a more rapid and certain repair.

Under ordinary circumstances no special constitutional treatment is necessary beyond that of securing good hygienic surroundings, easily digested food, restricted at first, and free action of the bowels. If there is great pain, opiates may be necessary, but here, as in other surgical indications for anodynes, a minimum amount should only be given. Usually rest, elevation, and relief of tension will be of greater benefit than opiates. If there is great restlessness, a bromide may suffice; if marked insomnia, one of the ordinary hypnotics. Great restlessness, with excitement and occasional delirium, without special evidence of pain or infective process, must call attention to the possible development of delirium tremens from a relatively slight injury (such as a crushed toe or a simple fracture), as it may precipitate an attack in one who has been a steady drinker, though perhaps not an excessive one. In such cases, in addition to the ordinary therapeutic remedies, the regular administration of whiskey should be advised.

TOXEMIA, SEPTICEMIA, SAPREMIA, PYEMIA

Toxemia applies to the diseases in which one or more poisons are present in the blood which are not necessarily of parasitic origin and production.

The word poisons is here used in a broad sense to cover any substance applied to the body, ingested, or developed within the body which causes disease. It of course includes ptomains, leukomains, toxins and sepsins.

Toxemia, according to this definition, would include the diseases due to poisons not arising from parasitic invasion of the tissues and fluids of the body, at times of vegetable and alkaloidal nature, such as strychnine or morphine; of animal origin, such as the toxin of snake venom, the ptomains of milk or shell fish; then again a mineral such as arsenic or lead; and lastly the leukomains arising from disturbed excretion and perverted metabolism and grouped under such terms as intestinal or uremic poisoning.

Septicemia may be defined as an acute febrile affection, characterized by marked nervous, cutaneous and visceral manifestations, and due to the introduction into the system of bacteria and their toxins from an infected wound. It applies to diseases which present poisons in the blood that are of parasitic origin, the parasite itself being either present or absent in the blood. Septicemia, in strong contrast to the definition of toxemia, would include diseases arising from the invasion of the tissues and fluids of the body by animal or vegetable parasites or their poisonous products.

Symptoms. The onset, as a rule, is slow, beginning from 4 to 7 days after an injury, with a chill, which is followed by fever, at first moderate, but soon becoming high. The fever presents morning remissions and evening exacerbations and may occasionally show an intermission. When the remission begins, there is a copious sweat. The pulse is small, weak, very frequent, and compressible; the tongue is dry and brown with a red tip; the vomiting is frequent, and diarrhea is the rule; delirium alternates with stupor, and coma is usual before death; prostration is very great, and visceral congestion occurs; the spleen is enlarged, ecchymoses and petechiae are noted, secretions dry up, urinary secretion is scanty or is suppressed, and the wound becomes dry and brown.

Blood examination detects disintegration of red globules and marked leukocytosis. When a wound becomes septic, red lines of lymphangitis are seen about it and there is enlargement of the related lymphatic glands. No thrombi or emboli exist in septicemia.

The prognosis is bad, and in some malignant cases death occurs within 24 hours.

Treatment is the same as for septic intoxication (see "*sapremia*"). Antistreptococci serum can be used, but the value of this method is doubtful.

Sapremia may be defined as an intoxication due to the absorption of dead saprophytes and their products (ptomains and toxalbumins).

Symptoms. The disease sometimes begins with a chill, followed by a marked rise in the temperature, but in most cases the latter is the first evidence of the disease. The skin becomes cold and clammy, there is marked prostration and sometimes diarrhea. When these manifestations occur while a wound is present, they are ominous, and the dangerous complications can be avoided if the dressing of the wound is renewed and perfect antiseptic precautions are taken to thoroughly remove all septic matter from its surface. The constitutional symptoms often disappear of their own accord, when the above has been done, unless the systemic intoxication has not already advanced to thwart all endeavors. There is also a diminution or suppression of the urine, and a blood examination shows leukocytosis.

Treatment. The treatment is at once to drain and asepticize the putrid area and to give large amounts of alcohol. Strychnine and digitalis are useful. Purge the patient, and favor diaphoresis, using in some cases the hot bath. Establish the action of the kidneys; allay vomiting by champagne, cracked ice, calomel, cocain or bismuth. Give liquid food every three hours. Feed on milk, milk and lime water, liquid beef, peptonoids, and other concentrated foods. Use quinine in stimulant doses. Antipyretics are useless. Watch for visceral congestion and treat it at once.

The use of saline fluid by hypodermoclysis or by venous infusion dilutes the poison and stimulates the heart, skin, and kidneys to activity.

In sapremia the blood contains the toxins and dead saprophytic organisms. In septicemia the blood contains both pyogenic toxins and multiplying pyogenic organisms. In sapremia the causative condition is putrid material lodged like a foreign body in the tissues. In septic infection the tissues themselves are suppurating, and both bacteria and toxins are absorbed by the lymphatics. Of course, septic infection may be associated with septic intoxication or may follow it. The symptoms of sapremia depend upon the amount of intoxication.

In septic infection, or septicemia, only a small number of organisms may get into the blood, but they multiply rapidly. A drop of blood from a man with septic infection will reproduce the disease when injected into the blood of an animal; hence it is a true infective disease. The wound in such a case is often small, and is commonly punctured or lacerated.

Pyemia may be defined as a condition in which metastatic abscesses arise as a result of the existence of pyogenic bacteria in the circulating blood, either free or contained in pus cells or thrombi.

Symptoms. The symptoms of pyemia are a febrile movement with a severe chill and a sudden marked rise in the temperature which lasts for a few hours and passes off with profuse sweating. The chills recur every other day, every day, or oftener. The general symptoms of vomiting, wasting, etc., resemble those of septicemia.

The lodgment of emboli produces symptoms whose nature depends upon the organ involved. If in the lungs, there is shortness of breath and cough, with slight physical signs.

In a suspected case of pyemia, always look for a wound, and if this does not exist, remember that the infection may arise from an osteomyelitis.

Chronic pyemia may last for months; acute pyemia may prove fatal in a few days.

Treatment. The treatment is the usual supporting one that should be employed in septic affections, and all suppurating focci must be opened and drained as soon as detected. Every branch of the irregular cavities must be opened and drained at the most dependent part, and the sinuses must be treated to prevent pocketing. Serum therapy is also indicated.

CHAPTER VI

HEMORRHAGE

Definition. The escape of blood from the blood vessels in great or small quantities, is called hemorrhage, and may occur either spontaneously or because of injury.

Spontaneous hemorrhage occurs in the organs and cavities of the body as a result of constitutional diseases, such as tuberculosis, syphilis, cancer, etc., in which erosion of tissue extends into vessels. It is also a result of a constitutional tendency. Persons with this, so called hemorrhagic diathesis, are known as hemophiliacs.

In hemophilia, uncontrollable bleeding may occur from trifling injuries.

Hemorrhage due to Injury may be classified as follows:

a—arterial

b—venous

c—capillary

(*a*) Arterial hemorrhage may be recognized by rapid, spurting jets of red blood, occurring synchronous with the heart beat.

(*b*) Venous bleeding (from a vein) occurs as a steady even stream of dark blood, not affected by the heart beat.

(*c*) Capillary hemorrhage is in the form of a steady stream oozing from the raw surface of a tissue. The color is intermediary, as both arterial and venous capillaries contribute to it.

Nature's Efforts to Control Hemorrhage. When an artery is severed, the inner and middle coats immediately retract and curl up within the lumen, partially closing up the cut end.

Blood has the property of clotting, if it comes in contact with anything but the natural endothelial lining of the vessels.

The curling in of the inner and middle coats retards the escaping stream and facilitates coagulation within the cut end of the vessel now formed by the outer coat alone. When the hemorrhage is severe, these processes are reinforced by an increased tendency to coagulate, and by a weakened heart action.

The Control of Hemorrhage. The object of treatment in every case is to check the flow of blood, and, though death from ordinary wounds is rare, yet the loss of much blood is weakening for a long time.

The principle on which we act in our efforts to permanently stop bleeding, depends on the power which the blood has of clotting, or as it is called, coagulating.

If by any means the blood can be made to "stand still" in a blood vessel at the point of injury, it will clot, thus forming a plug which prevents further escape.

In wounds involving only small veins or capillaries from which there is no distinct jet of blood (capillary hemorrhage), pressure of the thumb, a wad of sterile gauze intervening, will usually suffice in a few minutes. Gauze dipped in hot water applied to such wounds, also at times effects a stoppage of such bleeding. Often only tight bandaging is necessary.

Bleeding from large arteries or veins can be controlled temporarily by pressure directly over the wound.

Temporary control may also be obtained by digital pressure above or below the wound, if in a leg or arm, depending upon whether the escape is chiefly from a vein or an artery, for in any wound some of the bleeding will be capillary. This method, or the application of a tourniquet, will absolutely control bleeding in an extremity.

The pressure in arterial hemorrhage must be applied at a point nearer the heart and in venous hemorrhage at a point away from the heart.

A tourniquet may be devised from a handkerchief, a piece of rope or of rubber tubing wound around the limb and tightened just enough to arrest the main stream; in addition, pressure exerted over the wound will control whatever hemorrhage persists. Such a control can only be temporary, as the arrest of circulation in an extremity below the tourniquet for more than an hour or two might cause gangrene. However, there is no great fear of this occurring, as some blood reaches the parts through deep vessels.

Permanent control of such hemorrhages can only be effected by grasping the severed vessels in the open wound with artery clamps, and then ligating below the clamps with cat gut.

Deep-seated hemorrhages, in the abdomen or chest, can often be controlled by pressure directly over the wound until an open operation can be performed.

Deep pressure, with the fist upon the abdomen just to the left of the vertebral column, will compress the aorta and greatly reduce the escape of blood from any artery supplied by the descending aorta.

Hemorrhage in Chiropody. For the chiropodist, bleeding is an annoying and especially perplexing occurrence. The feet are the most bacteria-laden part of the body; here are warmth and moisture, congenial to bacteria, and a thick epidermis for their safe concealment. When hemorrhage occurs, therefore, its proper control along antiseptic lines is imperative.

The vessels severed are rarely of sufficient size to cause the escape of blood in an actual stream, but rather as a rapid oozing. It is, as a rule, capillary hemorrhage.

The methods for its control have already been described in this chapter, and will always stop such bleeding.

In chiropodial practice, however, the degree of bleeding determines the method of treatment, and, though the extreme may fall short of actual danger, it still behooves the operator to control it absolutely before dismissing his patient.

Easily Controlled Bleeding. The degree of bleeding or slight oozing, as it should be termed, incident to skiving a calloused surface, is well controlled with styptics.

In employing these substances it should be borne in mind that they are not usually antiseptic but, on the contrary, may harbor organisms which may be transferred to the wound and cause infection. The subsulphate of iron, commonly employed in the form of Monsel's solution, is usually employed because of its efficiency as a styptic, and because of the fact that it is less irritating than others. It, however, is not antiseptic and should be kept sterile and uncontaminated by dropping it upon the wound directly from the bottle, rather than by dipping the cotton-wound applicator into it, as is so frequently done. Even this does not prevent an originally sterile bottle of solution from becoming contaminated, exposure to the air, when the stopper is removed, admitting many bacteria each time.

A superior styptic has been supplied in the form of dry subsulphate of iron fused to small sticks of wood. These are efficient because of their cleanliness, each being used but once and at no appreciable expense.

It is needless to say that the dressing of even so slight a wound should prevent the admission of infection to the thousands of portals of infection which are present. A bandage is not indicated nor justifiable, and the cotton collodion cocoon suffices.

Persistent Bleeding. When bleeding occurs which does not yield to the effects of a styptic because of its constant washing away when applied, it becomes necessary to apply pressure to the wound. Frequently a wad of cotton or gauze, pressed firmly upon the bleeding area, will almost stop the bleeding in a few minutes, after which it becomes possible to apply the styptic. Should this, however, be found impossible and the bleeding resume when the pressure is released, clotting in the vessel can only be expected by the agency of either ligation of the tissue or any individual vessel or more commonly by tight bandaging. The latter procedure usually accomplishes the control of the hemorrhage incident to a deep dissection for papilloma or verucca.

A pad of several thicknesses of sterile gauze is placed upon the wound and held in place by a few turns of narrow bandage, applied quite tightly. Though blood may be seen to "spot" through this dressing, it should occasion no alarm unless the hemorrhage has been clearly either venous or arterial. Under such circumstances the spurting, either constant or intermittent, will give immediate evidence of its character. Active hemorrhage of this nature may yield to tight bandaging, but ligation of the vessel should be done.

Venous or Arterial Bleeding requiring ligation may be easily dealt with, and every chiropodist should be equipped with a small artery clamp with which to grasp the tissues; he should also be provided with sterile catgut, sizes 0 or 00, with which to ligate a bleeding vessel.

Antiseptic Precautions. In dealing with hemorrhage of even the slightest degree, it should be remembered that portals of entrance for bacteria upon the feet require every antiseptic precaution, both as to the treatment of the wound, and as to the instruments and dressings which come in contact with it.

For open wounds the U. S. P. tincture of iodin, diluted in water to one-half strength, is antiseptic and not extremely irritating.

Instruments dipped in pure phenol and dried on sterile gauze are rendered sterile and may be safely employed.

Dry sterile gauze in the dressing of a clean surgical wound is all that is necessary. Healing in the absence of infection will be prompt. The habitual use of ointments and wet dressings should be discountenanced, except in the presence of a real indication.

CHAPTER VII

BURNS, FROST BITE, ETC.

Among the causes of burns are: steam; hot water; melted glass, wax, rubber, sugar; molten metal; red-hot metal; gas and flame; burning wood, paper, clothing; electricity; X-ray; ultra-violet ray; chemicals; acid sulphuric, trichloracetic acid, common lye; alkalis; carbolic acid; iodin; croton oil, mustard, cantharides.

From these various causes there is very little difference in symptoms, course, pathology, and treatment. The molten lead burns are usually small in area, but of the third degree. The underlying tissues are often devitalized, especially around the feet, making a deep, pale, slow-healing ulcer. The same is true of many burns from electricity. The effects of X-ray burns are only seen after several days or weeks and stubbornly resist treatment. Ultra-violet ray burns may not show any effects at first, but develop symptoms in about six hours, sometimes accompanied by great pain. Such burns may be due to sunburn or powerful electric light.

The epidermis contains no blood vessels, but the mucous layer has lymph spaces between the cells, draining into the lymph spaces and channels of the dermis. Nowhere in the body are nerves more abundant than in the skin. Here we have nerves of motion to the muscles of the skin; nerves of pain, temperature, and touch; forming an intricate plexus of nonmedulated fibres sending their branches upward into each papilla, and even to the mucous layer of the epidermis. Vasomotor nerves supply the coats of most blood vessels of the skin, and trophic nerves are everywhere controlling the nutrition of each part. When it is considered what a complex organ the skin really is; how delicately its parts are adjusted to the body; how extremely sensitive its nerve supply, slight stimuli bringing responses and causing reflex action in far distant organs; how many the uses of the skin (protection, excretion, expression, and sensation in various forms), it can readily be understood how great is its importance, and the far-reaching results of its serious injury.

Burns are classified into three degrees: first, second and third. In every burn there are two layers of tissue to be considered: *first*, the layer destroyed—the dead flesh; *second,* the layer injured—the sick flesh.

BURNS OF FIRST DEGREE

Pathology.

(1) Destruction of the cells of the horny layer.

(2) Injury of the cells of the mucous layer with an excess of lymph. No blistering.

(3) Congestion of the subpapillary plexus with some destruction of the hemoglobin.

(4) Closing of the ducts of the sweat and oil glands.

(5) Slight edema of the underlying dermis.

Clinical Stages.

1st stage—hyperemia and pain.

2nd stage—edema.

3rd stage—peeling and staining the skin.

4th stage—cells of the horny layer replaced by pushing upward of cells from stratum lucidum.

BURNS OF THE SECOND DEGREE

Pathology.

(1) Destruction of cells of horny layer and sometimes of the germinal layer.

(2) Great exudation of fluid composed of lymph, fibrin, and broken-down cells in the lymph spaces of the mucous layer, forming blisters.

(3) Intense swelling and congestion of the papillary layer.

(4) Swelling of the connective tissue and elastic fibres in the true skin.

(5) Thrombosis in some superficial blood vessels.

(6) Leucocytes poured out around the blood vessels.

Clinical Stages.

(1) Stage of blistering, edema, dermatitis, toxemia, pain, chill and shock.

(2) Discharge or absorption of contents of the blister with shedding of dead layers of epidermis.

(3) Reproduction of cells of the mucous layer from those of the germinal layer, which have formed the floor of the blister.

BURNS OF THE THIRD DEGREE

Pathology. Charring of the whole skin through the reticular layer, or deeper. It may involve only skin, or include any underlying structures, fascia, muscles, blood vessels or bone. The essential feature is the total death of hair follicles, oil and sweat glands, with consequent destruction of all germinal epithelium.

Clinical Stages. (1) Stage of destruction of tissue with underlying inflammation. If extensive, this degree of burn causes shock, probably non-toxic. During the early stage there is apt to be great pain from injury to the nerves in the sick layer, but not so great as in that of second degree burns where the number of injured nerves is greater.

(2) The general effects (toxemia, blood changes, embolism, congestion of vital organs with resultant chill and shock) are probably little different from those in extensive burns of the second degree, as few burns are purely third degree burns, but if extensive they have also large areas of second degree burns.

(3) Stage of sloughing. During this stage the second degree portion of the burn passes through its various stages and heals. The dead tissue shows at its edges a line of cleavage from the surrounding

living skin. The slough is usually slow in coming away, owing to the direction of the connective tissue and elastic fibres which bind it to the underlying structures. This last stage lasts from one to three weeks. The process is more rapid in infected burns and the depth of this burn will depend upon the degree of heat to which the part was subjected, the length of time the heat was applied, and several other factors. The danger of infection is always great owing to: (a) presence of dead tissue; (b) the low resistance of adjacent sick tissue; (c) the open veins and lymph channels; (d) the adjoining skin which is difficult to sterilize; (e) the discharge of a large amount of serum which forms an excellent culture medium. There may be also severe hemorrhage as in any sloughing wound. The danger of this is greatly increased by infection, which breaks down the thrombi in the veins and arteries.

Stages of Granulation. The cavity left by the slough rapidly fills with new granulations. These have a tendency to rise above the surrounding skin.

Stage of Epidermis Covering. If skin grafting is not done, the new epithelium can be renewed only from the edges—a slow process often requiring months to cover the whole surface. Coincident with this stage is the stage of cicatrization. The granulations which fill the space left by the slough soon begin to contract—nature's effort to fill the gap. The granulations are irregular and abundant and for this reason the scar resulting from a burn is irregular, uneven, inelastic, contracted, distorted, protuberant and disfiguring.

Duration. First degree burns get well in a few days; those of second degree, in about from seven to fourteen days, and the healing of the third degree burns depends upon their extent and depth, severe ones requiring a very long time. As to scarring in a burn of the third degree, you can always predict it, although this can be minimized by early skin grafting.

Treatment. The local treatment is to be directed toward the limitation of the resulting inflammation; the prevention of septic infection; assisting the normal elimination of the eschar; the development of granulations and limitations of the deformity.

In burns of the first degree little or no treatment may be requisite; a mild dusting powder such as boric acid or sodium bicarbonate may be used, or picric acid in the strength of from half to one per cent.; a 5 per cent. boric acid ointment is also to be recommended.

Burns of the second and third degree require a different treatment. Suppose we are called to treat a severe burn of the second or third degree and find the patient suffering agonizing pain with oncoming shock and a chill. At once administer a hypodermic of one quarter to one half a grain of morphine; 1-40, to 1-20 grain of strychnine; and 1 to 1-100 or 1 to 1-50 grain of atropine. To stop the pain and combat shock, have the room warm, clear it of unnecessary furniture; order hot water bottles, and, if necessary, give a hypodermoclysis or a Murphy enema.

In a severe burn three things are more important than the local treatment: (1) to stop the pain; (2) to combat shock; (3) to provide for dilution and elimination of the toxins, which are thrown into the blood.

After having carried out the instructions given above, then proceed to do the local dressing. The clothing should be carefully cut away—never pulled off, or dragged over the burned area. A burn is at first sterile, and we must try to keep it so. Unless we believe that it has become infected through dirty handling, or by having had dirty clothing dragged over it, or a dirty blanket laid on it, it is best not to wash the burn. Pieces of gauze of necessary size are now spread thickly with an ointment and applied somewhat beyond the burned areas; over this cotton, and over all a bandage.

The patient is now put to bed, and if shock continues, the normal salt solution is repeated every eight hours and the patient is given plenty of water to drink.

Nourishment for the first three days should be liquid, on account of the intense congestion of the alimentary tract Food is gradually increased according to conditions. There should be the usual care of the bowels, skin and kidneys, but in our zeal over the local treatment, we should not forget that we have to care for a patient whose blood is loaded with toxins, and whose lungs, stomach, kidneys, and other organs are congested and filled with emboli. At first, dress the burns daily, gently wiping away the discharge of serum and broken down cells, which is poisonous and irritating, with dry gauze or cotton. Blisters are opened and pieces of loose skin removed with sterile scissors or forceps, but all skin is left in place as long as possible to protect the underlying, new forming skin. Every dressing should be made with a septic care: clean hands, clean gauze and clean instruments. As soon as the slough begins to form, if there is much odor, it is well to apply a continuous wet dressing. In case of a burn caused by carbolic acid, the skin is neutralized by the use of absolute alcohol (95 per cent.). In burns from trichloracetic acid, use alkaline remedies as sat. sol. of sodium bicarbonate. Burns from caustic alkalies are neutralized by vinegar or by some other mild acid such as boric acid. A so-called X-ray burn is not a burn at all; the observable results of such an accident are not manifested until several days or even several weeks after the application of the rays, at which period an inflammatory or a gangrenous process arises, which begins within the deeper tissues and subsequently involves the surface. These burns are often accompanied by loss of hair or of nails in the damaged area; they frequently remain unhealed for months; if they heal at all, they are very painful, and are not improved by the treatment which relieves ordinary burns. In some cases the consequences are very serious. Ambrine is a newly proposed remedy.

Effects of Cold. The more serious effects consequent upon exposure to sudden or prolonged cold are termed *frost bite*. In this condition the feet are commonly affected, and very often the freezing is so complete that upon thawing, the parts are found to be absolutely dead or their vitality so impaired by the cold that after reaction, strangulation and inflammation of the tissues occur, producing gangrene. As in burns there are three degrees of freezing, viz., first, second and third. In the first, the redness, numbness and tingling which follow exposure to intense cold are succeeded by loss of power, usually commencing in the toes, and loss of sensation, the parts becoming anemic and cold. In the second degree the skin is red or bluish and is covered by blebs with clear hemorrhagic contents. If the epidermis only is lifted up there is quick, scarless healing, but in the majority of cases the deeper tissues are involved. In frost bites of the third degree there are blebs and crusts which eventually mortify. Parts hopelessly frozen are at first anemic, cold and insensible but after reaction sets in they become swollen and discolored or they shrivel up and contract. It is not unusual for the part to show no change for some days and then to become blue or black; a line of demarcation forms and the dead tissue sloughs off.

Treatment. Reaction must be gradual. The room should be of low temperature; the affected part should be immersed in ice water; gentle friction or rubbing lightly with snow is oftimes efficacious. When the temperature is normal, stimulating friction with soap liniment, alcohol, and water and spirits of camphor with elevation of the parts, is advisable. The room may be gradually warmed and the parts exposed should then be covered with cotton. As reaction progresses warm, stimulant drinks may be cautiously administered. If excessive reaction takes place, evaporating lotions of alcohol and water may be used. Where a large surface is frozen, prolonged immersion in a bath may be employed after reaction has been established. When gangrene is present, surgical intervention is imperative.

Chilblain occurs in individuals with a feeble circulation or in the anemic or strumous, though healthy young people are not immune. The feet are very often attacked, especially the heel and the borders of the feet, but any of the peripheral parts may be affected. The areas are bluish or purplish red, swollen, cold to the touch, tender, itching and burning. Neglect and friction will produce severer grades of inflammation, with vesicles, bullae, pustules and ulceration or even gangrene, with or without the formation of bullae. There may be a favorable termination or fatal septicema may supervene.

Treatment. This should be preventive by protecting the feet, wearing warm clothing, by exercise, and the administration of tonics. Local immersion of the affected part in hot saturated solution of alum relieves the venous congestion and the itching. In severe cases, heating too rapidly, or overheating, should be prevented so as not to restore a too rapid reaction. A strong faradic current, ten minutes thrice daily, or the electric bath, ten to fifteen minutes daily, is beneficial. In ordinary cases, balsam of Peru or 10 per cent. ichthyol ointment, rubbed in, is all that is required. When there is ulceration, antiseptic dressings should be applied.

CHAPTER VIII

FISTULAE; FISSURES; SINUSES; ABSCESSES; FURUNCLES; ULCERS

A **Fistula** (pl. fistulae) is an abnormal communication between the surface and an internal part of the body, or between two natural cavities or canals. The first form is seen in a rectal fistula, the second in vesicovaginal fistula. Fistulae may result from a congenital defect and can arise from sloughing, traumatism and suppuration. Fistulae are named from their situation and communication.

A **Fissure** is a crack and in podiatry, has special reference to a condition found in the toeweb.

A **Sinus** is a tortuous track opening usually upon a free surface and leading down into the cavity of an imperfectly healed abscess. A sinus may be an unhealed portion of a wound. Many sinuses may be due to pus, burrowing subcutaneously. A sinus fails to heal because of the presence of some irritant fluid (as saliva, urine) or, because of the existence of some foreign body, as dead bone, a bit of wood, a bullet, a septic ligature, or because of rigidity of the sinus wall, which rigidity will not permit collapse. The walls of a tubercular sinus are lined with a material identical with the pyogenic membrane of a cold abscess. Sinuses may be maintained by want of rest (muscular movements) and by general ill-health.

Treatment. In treating a fistula, remove any foreign body; lay the channel open, curet, touch with pure carbolic acid, and pack with iodoform gauze. In obstinate cases, entirely extirpate the fibrous walls; sew the deeper parts of the wound with buried catgut sutures, and approximate the skin surfaces with interrupted sutures of silkworm gut. Fresh air is necessary; nutritious food and tonics must be ordered.

Acute Abscesses. An abscess may be defined as a circumscribed cavity of new formation, containing pus. An essential part of this definition is the assertion that the pus is in a cavity of new formation; is an abnormal cavity; hence pus in a natural cavity (pleural or synovial) constitutes a purulent effusion, and not an abscess, unless it is encysted in these localities by walls formed of inflammatory tissue.

An acute abscess is due to the deposition and multiplication of pyogenic bacteria in the tissues or in inflammatory exudates.

When abscesses form in an internal organ or in some structure which is not loose like connective tissue, for instance, in a lymphatic gland, a mass of pyogenic bacteria floating in the blood or lymph, lodges, and these bacteria, by means of irritant products, cause coagulation necrosis of the adjacent tissue and inflammatory exudation around it. The area of coagulation necrosis becomes filled with white blood cells, and the dry necrosed part is liquefied by the cocci. Suppuration in dense structures causes considerable masses of tissue to die and to be cast off, and these masses float in the pus.

An abscess heals by the collapse of its walls, and the formation of an abundance of granulation tissue; in many cases granulations of one wall join those of the other side, the entire mass of granulations being converted into fibrous tissue, and this tissue contracting, heals by third intention. If the walls do not collapse, the abscess heals by second intention.

Symptoms. The symptoms of an acute abscess may be divided into (1) local, (2) constitutional. Locally there is intensification of inflammatory signs; swelling enormously increases; the discoloration becomes dusky; the pain becomes throbbing, and the sense of tension increases; the cutaneous surface is seen to be polished and edematous, and after a time, pointing is observed and fluctuation can be detected. The constitutional symptoms are usually limited to chills and fever, depending upon the severity of the infection.

Treatment is free incision and drainage. The wound should be opened early, if possible even before pointing or fluctuation, to prevent destruction, subfascial burrowing, and general contamination; drainage is continued until the discharge becomes scanty, thin and seropurulent.

Chronic Abscess is a term referring only to time. Usually a tubercular abscess is designated as a chronic, cold, or scrofulous abscess. It is an area of disease produced by the action of the tubercular bacilli and is circumscribed by a distinct membrane. The symptoms present no inflammatory signs. Constitutional symptoms are trivial or absent unless secondary infection occurs. The treatment of these cold abscesses depends upon their location.

A Furuncle or Boil is an acute and circumscribed inflammation of the deep layer of the skin and the subcutaneous cellular tissue, following on bacterial infection of the hair follicle through a slight wound (by scratching, shaving), with the staphylococcus pyogenes aureus.

Symptoms. The symptoms of a boil are as follows: a red elevation appears, which stings and itches; this elevation enlarges and becomes dusky in color, a pustule forms that ruptures and gives out a very little discharge which forms a crust; inflammatory infiltration of adjacent connective tissue advances rapidly, and the boil in about three days consists of a large red, tender, and painful base, capped by a pustule and some crusted discharge. In rare instances, at this stage, absorption occurs, but in most cases the swelling increases, the discoloration becomes dusky, the skin becomes edematous, the pain severe, and the centre of the boil becomes raised. About the seventh day rupture occurs, pus runs out, and a core of necrosed tissue is found in the centre of a ragged opening. The hair follicle and the sebaceous gland, which have undergone necrosis, are found in this core. Healing by granulation will occur; the constitution often shows reaction during the progress of a boil.

Boils may be either single or multiple, and the development of one boil after another, or the formation of several boils at once, is known as *furunculosis*.

Treatment. The treatment consists of crucial incision and the application of a wet dressing.

An Ulcer may be defined as the loss of substance due to necrosis of a superficial structure, and the causes of ulcers may be divided into (1) predisposing and (2) exciting. In the former, age, sex, occupation and social condition have to be considered. The exciting causes are traumatism and infection.

The chief varieties of ulcers seen on the leg and foot are as follows: indolent or callous; varicose; tubercular; syphilitic; epitheliomatous; diabetic; perforating and blastomycotic

In indolent or callous ulcer, the cause may be divided into general and local. Among the former may be mentioned typhoid fever, chronic nephritis, anemia, poor hygiene, improper food, overwork, and lack of sleep. Local causes: old scar tissue, extremes of heat or cold, irritation of the tissues, injury, the presence of a foreign body such as dead bone, splinter, etc.

Symptoms. The most common location of these callous ulcers is on the inner side of the lower third of the leg. They show a great variety in size, shape, appearance and base, edges and surrounding area, and in accordance with these differences, many different names are applied to them. The size varies from a small ulcer less than one centimeter in diameter, sometimes found with varicose veins, to the large ulcerations which surround the leg and are called *annular* ulcers. The shape may be round, very irregular, or funnel shaped. The base may be much or slightly depressed, or the granulations may be at a higher level than the surrounding edges. When the granulations are large, irregular, and bleed easily, they are spoken of as *exuberant*; when pale, soft and flabby, as *weak* or *edematous*; when small and slowly growing, as *indolent*.

A peculiarly painful form of chronic ulcer is found over the internal malleolus, and most frequently in women of middle age; it is often associated with menstrual disorders and is known as a*congested* or *irritable* ulcer. It begins as a small area of congestion over the internal malleolus, which gradually increases in size and becomes dark and more dusky in the centre, due to the deposit of blood pigment caused by chronic congestion. The skin next becomes hard, dry, scaly and pigmented, while the subcutaneous tissues lose their elasticity, becoming inflexible, hard and adherent to the deeper structures. Then, as a result of slight traumatism or even without injury, the centre of the area breaks down and an ulcer develops. It may be circular or irregular in shape and may be quite deep or superficial. The edges are sharply cut, and both base and edges are bound down to the deeper tissues. The intense pain of the ulcers is supposed to be due to pressure upon the terminal nerve filaments in the dense sclerotic tissue. This form of ulcer is very often difficult to cure and shows a tendency to return after healing.

Treatment. This naturally depends upon the time the ulcer is seen and the conditions present. If there is considerable inflammation, accompanied by marked cellulitis and pain, the milder wet dressings, such as boric acid or Thiersch are indicated. Rest, of course, is the most important factor. The patient must be prohibited from walking, and if necessary, the movements of the neighboring joints must be prevented by the application of suitable splints. After the acute inflammatory symptoms have subsided the granulations must be stimulated, (see Chapter XIX).

Varicose Ulcer. To chronic ulcers of the leg associated with varicose veins, especially of the smaller venous radicles, the name varicose ulcer has been given.

Symptoms. The usual development of this variety of ulcer is as follows: persons who suffer from varices of the leg usually complain for some time before the external manifestation of the disease, of a deep aching pain in the limb, with a sense of weight, fullness, and fatigue. In a more advanced state of the disease, the ankles swell after a day's hard work, and the feet are constantly cold; an embarrassed state of the circulation is denoted by these symptoms and the deep seated veins begin to swell. After a time, which varies with the idiosyncrasy and occupation of the patient, small soft, blue tumors are seen at different points of the leg, most of them disappearing on pressure, but returning when this pressure is removed or when the patient stands up. Each little tumor is caused by a vein dilated at the point at which it is joined by the intramuscular branch. Around many of these tumors a number of minor vessels of a dark purple color are clustered, these being the small superficial veins which enter the dilating vein and in which the varicose ulcer is often of a brownish blue color, due to a deposit of pigment. Frequently a leg, which is the seat of varicose veins, or which is edematous from other causes, is attacked by acute eczema. The recognition of varicose ulcers is usually easy but the mere presence of enlarged veins, it should be noted, is not pathognomonic, because they may often exist along with ulcers of other origins, tuberculous, syphilitic, etc.

The surface of varicose ulcers usually presents imperfect and unhealthy granulations, secreting a more or less thin and offensive pus, and the granulations are sometimes covered with membranous exudation. The edges and base are thickened and callous, and enlarged veins, capillary or otherwise, are present near the circumference and often amount to genuine blood tissue which tunnels the infiltrated tissues. In examining such an ulcer one gets the impression of a great pigmented scar, the centre of which has broken down.

Lymphangitis and venous thrombosis are not of infrequent occurrence in connection with varicose ulcers, while embolism and even pyemia are sometimes in evidence. Among the most frequent complications is cellulitis, and this may sometimes be so severe as to necessitate operation. Erysipelas may also occur in cases of varicose ulcer, and hemorrhage is a common and serious complication and has at times been fatal.

Differential Diagnosis

CALLOUS	VARICOSE	SYPHILITIC
History:		
injury	varicose veins or phlebitis.	syphilis.
Situation:		
where the injury occurred.	usually in lower third of leg.	usually upper third of leg, posterior aspect.
Base:		
shallow, inflamed, often grayish yellow.	bluish, pigmented, granulations, sluggish, usually superficial.	dirty, sloughing, deep, often greenish in color.
Edges:		
not elevated or thickened.	undermined or thickened space, very irregular.	punched out thin and undermined shape, round or serpiginous.
Surrounding area:		
red and inflamed.	pigmented, varicose veins, often edema and eczema.	dusky red, scars of old syphilitic ulcers.

Healing:

rapid under antiseptic treatment.	support of veins, operate and remove veins.	mercury and iodides necessary, or neosalvarsan.

Treatment. The treatment of varicose ulcers must be based on antiseptic cleanliness, and the improvement of nutrition by improvement of the circulation of the blood and lymph. Then again the treatment will vary according to the time when the ulcer is first seen by the surgeon. In aggravated ulcers, especially those accompanied by crusts, foul smelling discharges and various inflammatory conditions, the leg should be washed once or twice daily with soap and water, cleansed with a piece of sterile gauze, and shaved when necessary. Warm applications should be employed such as Wright's solution, boric acid; Thiersch and the stronger antiseptics are uncalled for, as they often induce eczema. Under such treatment, in most cases, the swelling and irritation will subside and the ulcer will become clean and more healthy in appearance, especially if the patient be confined to bed with elevation of the limb. Rest always seems to the patient a useless waste of time, but in reality time is thus saved. It is by far the most important point in the treatment of ulcers of the leg in which poor circulation is a factor, but the plan must be carried out consistently in order to obtain the best results. The condition does not admit of occasionally walking about the house or of sitting in a chair. However, when circumstances do not permit of the recumbent position, the veins can be supported in various ways. Bandages of plain rubber, or rubber cloth, or cloth woven and rendered elastic by the character of mesh, or elastic stockings, or flannel, gauze, or muslin bandages, can be used. It is preferable to use flannel bandage (see Therapeutic measures) for the reasons mentioned. The best means of obtaining the support, however, is by the use of Unna's Paste. The technic and application of this method of treatment has also been described (Therapeutic measures).

Operations upon varicose veins are frequently called for in aggravated cases, provided the general condition of the patient permits. Briefly, these many consist in multiple ligations, in ligation of the internal saphenous alone, in extirpations of large or small sections of varices, in circumcision of the skin above the ulcer, or of the ulcer itself, tying all the veins and reuniting the cuticle. However, it must not be forgotten that in the presence of an ulcer, infection of an operative wound is likely to occur.

Syphilitic Ulcers may result from pustules or they may begin as tertiary sores. They occur frequently where the integument is thin or where the part is kept moist by the natural secretions. The deep ulcers of tertiary syphilis develop from gummata. These are variously sized deposits largely made up of large spheroidal cells and a few giant cells. They are poorly supplied with blood vessels and undergo coagulation necrosis, but do not tend to suppurate until infected. Sooner or later the overlying skin becomes involved, either with or without a pyogenic infection, and the gumma sloughs out leaving the typical syphilitic ulcer. A protozoa microbe (Schaudinn's and Hoffmann's organism) is now the recognized cause of syphilis. It is called the*spirochaeta pallida* or *treponema pallidum.*

Symptoms. When a syphilitic ulcer develops it usually assumes one of two types, superficial or deep. The former may appear comparatively early in the disease. It usually varies in size from a quarter to a half dollar piece, has a circular outline, sharply cut, indurated edges, and a dirty greenish base. The deep ulcers result from the breaking down of gummata. They are, at the beginning, surrounded by a reddened area of inflammation, the small ones being crater like, with punched out edges, the larger ones having overhanging, thin, soft, inflamed edges. The base is indurated, of a dusty red color and dirty or sloughing in appearance, the slough being often of a greenish color. The discharge is thin, frequently bloody, and contains debris from the broken down gumma. The surrounding skin is indurated, of a dusky red color and dirty or sloughing in for some time, they loose their characteristic appearance and take on the form of simple chronic ulcers. The scar remaining is characteristic. It is thin, of a dead white color, pigmented here and there, and when pinched it wrinkles like tissue paper. Thin form of syphilitic ulcer is found most frequently on the upper third of the leg. When ulcers are accompanied by enlarged veins, it is extremely difficult at times to make a differential diagnosis between a luetic ulcer and one of a varicose type. The chief differential points are as follows:

Location:

Varicose ulcers, the lower third of the leg.

Syphilitic ulcers, the middle and upper third of the leg.

Appearance:

Varicose, irregular, not undermined, granulations reddish.

Syphilitic, typical punched out edges, sharp, and undermined, greyish discharge, thin and watery.

Number:

Varicose usually single.

Syphilitic, multiple, having a tendency to coalesce and form one large ulcer.

A very important point to remember is that a syphilitic ulcer, once healed, usually remains so. At times it is extremely difficult, even in view of the different points already mentioned, to make a distinct diagnosis between a varicose and a syphilitic ulcer; then the Wasserman reaction should be resorted to, but too much stress should not be placed upon its findings. It may happen that a patient having a suspected luetic ulcer is given mercurial treatment with the result that the reaction is negative, but this should not exclude the possibility of syphilis existing. A positive Wasserman in a case of chronic ulcer with enlarged veins which refuses to heal, warrants a diagnosis of a syphilitic lesion. In a great many cases the Noguchi luetin skin reaction is of great aid in establishing a diagnosis.

Treatment. The treatment is both local and general. As regards local treatment, if the ulcer secretes freely, either the black wash or a solution of bichloride, varying from 1 to 5000 to 1 to 10000 should be employed. Where there is very little discharge, calomel powder is indicated. In addition, it is understood that a firm compression bandage be applied (especially in those cases complicated with enlarged veins) beginning at the base of the toes and carried up to the knee.

The general treatment consists of the intravenous injection of salvarsan or neosalvarsan (10 grains), or the intramuscular injection of bichloride of mercury, one quarter of a grain, or 10 minims of a 10 per cent. suspension of salicylate of mercury. In addition, mercurial rubs and the administration of iodides and mercury internally are advised.

A Tuberculous Ulcer usually results from the bursting through the skin of a tuberculous abscess. The base is, soft, pale and covered with feeble granulations, and gray shreddy sloughs. The edges are of a dull blue or purple color and gradually thin out toward their free margins, and in addition, are characteristically undermined, so that a probe can be passed for some distance between the floor of the ulcer and the thinned out borders. At times the edges are solid and puckered, being scarlike in character. Thin, devitalized tags of skin often stretch from side to side of the ulcer. The outline is irregular, small perforations often occur through the skin and a thin watery discharge containing shreds of tuberculous debris escapes. The ulcer is usually superficial and very little pain is present. At times it is crusted over, the crust being thin and of a brown or black color. Again it may be progressing at one point and healing at another. It is slow in advancing but often proves very destructive. The scars left by its healing are firm and corrugated, but are apt to break down.

Treatment. The local treatment calls for special mention. If the ulcer is of limited extent, the most satisfactory method is complete removal by means of the knife, scissors, or sharp spoon, of the ulcerated surface and of all of the infected area around it, so as to leave a healthy surface from which granulations may spring. If the raw surface left is likely to result in cicatricial contraction, skin grafting should be employed.

The general treatment should consist of tonics, plenty of fresh air, and a good nutritious diet. Bowels must be regulated.

Perforating Ulcer of the Foot occurs in connection with lowered resisting powers of the tissues, due usually to some lesion of the nerves or vessels. The ulcer is circular in shape, painless, with callous borders, and eats progressively into the deeper tissues and bones, and has little or no tendency to heal.

Etiology. Although formerly looked upon as a specific disease, perforating ulcer is now known to depend upon many local and general conditions of which it is occasionally a more or less accidental manifestation. The various theories as to its immediate causation may be divided into: (1) mechanical, (2) vascular, (3) nervous, (4) mixed.

The Mechanical Theory regards injury as the sole cause, due in most instances to the pressure or rubbing of a shoe. If this explanation were adequate, however, such ulcers would be extremely common, while in reality they are rare.

The Vascular theory assumes that arteriosclerosis is always present, and causes ischemic necrosis through arterial and capillary thrombosis.

The Nerve theory, which is the one most commonly accepted, is that perforating ulcer is always of trophic origin and depends upon a chronic peripheral neuritis. In support of this assertion, attention is called to certain interstitial and parenchymatous alterations frequently demonstrable in the nerves of the affected part. It must not be forgotten, however, that these nerve changes may be due to secondary disturbances in nutrition, depending upon arteriosclerosis as in senile, diabetic, and other forms of gangrene.

According to the Mixed Theory either vessels or nerves, or both may be at fault. It admits that traumatism is an important factor, although seldom if ever an exclusive cause. Perforating ulcer is observed in connection with various diseases and conditions, the most prominent of which are locomotor ataxia, fractures of the spine, injuries of the cord, diabetes, spina bifida, syringomyelitis and injury and division of the peripheral nerves. Perforating ulcer from lesions of the central nervous system is comparatively rare and it is doubtful if it is ever due to embolism or to ligation of the arteries.

The three most prominent causes, therefore are, (1) affections of the spinal cord (2) injuries of the peripheral nerves and (3) diabetes.

This variety of ulcer is seen more frequently in males than in females, and it is almost exclusively confined to adults, especially between the ages of forty and sixty. Occupations requiring standing or walking are strong predisposing causes, provided a tendency to the disease exists. A poor fitting shoe and deformities of the foot giving rise to excessive pressure or irritation, are of much importance in determining the appearance and location of the ulcer. It rarely appears in children, unless it is associated with spina bifida.

Symptoms. Perforating ulcer has a marked tendency to develop where pressure and irritation are greatest, which is almost always upon the sole of the foot at the junction of the great or little toe with the metatarsus. It may occur, however, upon the heel, the sides of the foot, the plantar surface of any portion of the great toe, or even upon the centre of the sole, these unusual situations being most commonly found associated with diabetes. When talipes or hammertoe exists, the ulcer is apt to occur wherever pressure is pronounced, even upon the dorsum of the foot or the ends of the toes. Usually but one foot is affected, although both feet may be involved, in which case the disease is termed symmetrical.

Three stages may be recognized in the development of the ulcer: (1) the formation of callosities, (2) superficial ulceration, (3) deep ulceration. Very frequently in tabes and in diabetes, a purulent blister is the first indication of trouble, but usually a marked epithelial thickening, in the form of a corn or a bunion, is the initial symptom. Sooner or later the centre of a callosity breaks down into a bluish, unhealthy, indolent, superficial ulcer, secreting a small quantity of watery pus, and with an offensive odor. The sore is circular as though punched out of the callous tissue, the latter at times so thickened and overhanging that the ulcer is almost concealed beneath it. There is little or no tendency to heal, even under exacting treatment, and if recovery should take place, a speedy relapse is the rule, even with the patient remaining in bed. The indolent and foul ulcer tends to eat deeply into the adjacent tissues, progressively involving bursae, tendons, muscles, joints, and bones. A deep round hole results, which may even perforate the foot. The most striking symptoms are chronicity, stubborn resistance to treatment, and the absence of pain and tenderness.

The fact that perforating ulcer is so often found in connection with lesions of the nervous system accounts for the abnormalities of sensation, motion and reflexes which accompany it. This explains the various trophic disturbances which are very often observed, such as epithelial growth, not only in the vicinity of the ulcer, but occasionally over the entire foot and leg; also eczema, erythema and excessive perspiration. The nails are frequently thickened and distorted and the subcutaneous cellular tissues are so changed as even to suggest elephantiasis. Inflammatory complications, sometimes serious, are not uncommon owing to infection through the ulcer, and an ascending neuritis may even result in myelitis. Gangrene from arteriosclerosis is also frequently seen.

Treatment in those predisposed to diabetes and tabes, deserves prophylaxis consideration. The shoes must fit accurately and without undue pressure; much walking is to be avoided; when ulceration has begun the recumbent position and cleanliness are of paramount importance. The callous epidermis should be removed so as to render the ulcer as superficial as possible. Dead bone must be scraped away or extracted, if in the form of a sequestrum, and drainage must be perfected by enlarging the opening. Sinuses should be enlarged and any pockets found should be thoroughly opened. It must be emphasized, however, that operative interference should be undertaken with care and discretion in order to avoid necrosis and infection. Periodic curettments and cauterizations with silver nitrate are often of benefit, as are also the employment of dry iodoform gauze as a packing, together with the occasional use of various moist dressings. Both the constant and interrupted currents of electricity have been resorted to with benefit, sometimes locally and sometimes applied to the spinal cord or affected nerves. Measures directed to the improvement of the circulation of the foot, such as massage, stimulating baths, and lotions, are of service.

Bier's Arterial Hyperemia, in the form of baking of the foot by means of a gas or electric apparatus, especially devised for the purpose (Tyrnauer) is of great benefit, more so when there is a neuritis accompanying the ulcer. The baking should be done once a day for from ten to twenty minutes, and the temperature should be gradually increased from 100°F. to 300°F., depending upon the patient's ability to tolerate heat.

The passive, venous or obstructive form of hyperemia is absolutely contraindicated in this class of ulcers. The initial cause of the trouble must receive attention, because upon its successful management depends the cure, much more so than upon the local measures.

Diabetics and syphilitics should receive appropriate treatment. The bad cases, especially where gangrene or serious infection exists, may require amputation, but unless this can be done in sound tissue with adequate innervation, a perforating ulcer may develop upon the area exposed to the pressure of an artificial limb. Resection of joints is usually of little benefit. The most satisfactory operative results in this class of ulcers have been obtained by stretching the posterior tibial nerve, together with scraping the ulcer, or, better, by excising it, followed by immediate suture of the wound. The operation is best done through a curved incision beneath the internal malleolus, the nerve being isolated and vigorously stretched in both directions by means of some blunt instrument inserted beneath it. Sometimes the external or internal plantar nerve alone is treated in this manner.

Blastomycotic Ulcer. This is not a common condition in the lower extremity. It is found near the lower third of the leg, and begins as a papule or papulo-pustule, soon becoming covered with a crust which, on removal, discloses a papillomatous area. The typical ulcer is elevated, verrucous or fungating, with a soft base which is infiltrated with a seropurulent secretion. The border is dark-red or purple and slopes more or less abruptly through the normal skin, from which it is sharply defined. The quickest and most positive method of differentiation is by means of the tissues. The organisms are fungi, known as the blastomycetes, saccharomyces or yeasts, characterized especially by their mode of multiplication or cell division, called budding.

Treatment. In all cases, thorough cleansing of the ulcer with antiseptic lotions, as previously described, is of great benefit. Complete extirpation of the ulcerative lesions has been successful, but curetting does not always prevent their recurrence. Potassium or sodium iodide in large doses (totaling from 100 to 400 grains per day) and radiotherapy seem to be the most efficacious forum of treatment. Copper sulphate in a 1 per cent. solution as a wash for external use and also in one quarter of a grain doses internally, has in some cases given good results.

Epitheliomatous Ulcer. In none of the more common ulcerative skin lesions would the conditions for the development of cancer seem to be more favorable than in chronic dermatitis with ulceration; the despised and neglected varicose ulcers of the leg. The extreme chronicity of the inflammatory process, often lasting for many years; the age of the patient, which is usually advanced; the almost inconceivable neglect of the lesion in many cases, so that the persistent presence of foul and decomposing secretion and of the products of tissue necrosis is common: the frequent absence of even an attempt at cure; the fact that most of these patients are compelled to be on their feet all day and thus keep up and increase the unfavorable conditions; and, finally the circumstance that in many of them the added history of alcoholism, of renal or cardiac disabilities, or of other chronic affections is also present; all of these factors would lead to the presumption that in this ulcerative lesion, above all others, carcinomatous degeneration would be the most common.

While so few instances of cancer secondary to varicose ulceration are seen, it rarely appears before the age of forty. It is usually seen where varicose ulcers as well as the scars they produce are found. The base of the characteristic ulcer is hard, nodular and irregular, made up of firm warty granulations, and often covered with sloughs. It bleeds easily and has a foul discharge. The edges are hard and everted. The borders and base present a peculiar and striking thickness and hardness, as though the ulcer were imbedded in cartilage, while the granulations feel firm and appear red and warty. The amount of pain, the involvement of neighboring lymphatic glands and the rate of growth vary. Epitheliomata which have developed from congenital warts, moles, or nevi are apt to be very malignant. When epitheliomatous degeneration occurs in a chronic ulcer, it first begins to get hard about the edges, which become everted and gradually bound down to the deeper tissues. The granulations about the margins become large, red, nodular, hard and bleed very readily. This condition spreads over the entire ulcer, which assumes a sloughing and foul character. The diagnosis is confirmed by the microscopic examination of a section cut from the edge of the ulcer.

Treatment. Malignant ulcer can be cured only by the destruction or removal of the new growth. For its treatment, caustics with or without curetting, excision or radiotherapy may be employed. The best caustics are arsenic, chloride of zinc, caustic potash and formalin.

The objections to this method are the extreme pain; the lack of certainty as to the removal of all of the neoplasm; the fact that the lymphatics and glands are not dealt with, as well as the fact that unless the treatment is thorough, the growth is stimulated rather than retarded. The scar is also apt to be unsightly. Without doubt excision forms the best method of treatment. The incision should be wide of the ulcer, and all indurated tissues and any lymphatics or glands that are involved must be removed.

In some cases it may be necessary even to amputate the leg in order to effect a cure. The X-rays from the Coolidge tube are to be recommended, as the cross fire effect of these rays in some cases is of great benefit. Recently radium has been used in these ulcers of the leg with good results. The gamma rays are to be preferred as they are more penetrating and should be applied two or three hours a day for a number of days. At least from 50 to 200 milligrams of radium bromide must be used in order to obtain any effect. Recently beta rays have been found to be as effective as the gamma rays. In order to prevent a radium burn the rays have to be filtered before they are applied.

CHAPTER IX

DISEASES OF JOINTS – THE SEROUS AND SYNOVIAL MEMBRANES

The moist glistening membrane lining the abdomen (*peritoneum*) and that lining the chest (*pleura*) are similar to the synovial sac between the bone ends at joints or the synovial sheaths of tendons.

Bursae. A bursa, which is a sac lined with serous membrane, placed over a joint or other prominent part for protection, is also quite similar. All of these membranes are smooth and moist, giving lubrication to movable parts, thus: the peritoneum covering the intestines, permits of their easy worm-like action within the abdomen; the pleura makes for the free rise and fall of the lungs; the *synovial sacs* of joints allow the bones to ride smoothly one upon the other; the *synovial sheath* of a tendon acts like a silken sleeve in which the tendon slides up and down and, lastly, pressure over a bony point causes the member to move aside because of the slipping of the walls of the bursa, one upon the other, when compressed.

INJURIES AND DISEASES OF BURSAE.

Synovial bursae exist normally in connection with tendons or with certain joints, and may be developed by continued friction or pressure at certain parts of the body. Deep bursae are sometimes connected with the joints, or are in very close relation with them.

Injuries of Bursae. Wounds of bursae may be either contused, incised, lacerated, or punctured, and, if they become infected, may prove most serious injuries. Wounds of bursae should be thoroughly disinfected and drained; they usually heal with obliteration of the sac.

Acute Bursitis. This affection usually results from an injury or from continuous irritation of a bursa, and is characterized by tenderness, pain, redness of the skin, and swelling or distension of the bursa. If suppuration occurs, the inflammation is apt to extend to the surrounding cellular tissue, or, if in close proximity to a joint, the latter may be involved. Bursitis can usually be diagnosed from other affections by the rapidity of development of the inflammatory symptoms, the location of the swelling in relation to certain tendons or joints, and its globular shape.

Treatment. This consists in elevating the part and putting it at rest on a splint, and in the application of cold or pressure. If, however, the pain and swelling due to effusion continue, and there is evidence of suppuration, the bursa should be freely opened and irrigated, and subsequently packed with sterilized or iodoform gauze. Under this treatment the cavity soon becomes obliterated as healing occurs. The bursae most commonly involved are the *prepatellar* and that over the metatarsal joint of the great toe.

Chronic Bursitis. This affection may result from acute bursitis which does not terminate in suppuration, or may develop slowly from long continued irritation or pressure, or from tubercular infection of the bursae and is accompanied by little pain.

The most marked feature in chronic bursitis is the distension of the sac with fluid, and in some cases the walls of the sac become so thickened that the bursa is converted into a solid tumor. Chronic bursitis of the prepatellar bursae is not infrequent, and is commonly known us *Housemaid's knee*, resulting from long continued pressure upon the knee occurring in those whose occupation causes them to constantly bear pressure upon this part.

Gumma of the prepatellar bursa is very common, and should be suspected in every case of suppuration of this bursa without assignable cause. It often results in extensive sloughing.

Hernial protrusion of a portion of a bursa is sometimes seen after injuries of bursae.

Treatment. The treatment of chronic bursitis, if the sac is distended with fluid, consists in removal of the fluid by aspiration, or by making an incision and introducing a drain. The greatest care should be observed to keep the wound aseptic. The bursae may be removed by dissection. This is the only treatment which is likely to be of use in cases where the bursa is very thick or is converted into a solid tumor. In removing these growths by dissection, great care should be exercised to avoid opening the neighboring joints.

Bunion. This is a bursal enlargement over the metatarsophalangeal articulation of the great toe, which is very frequently observed with hallux valgus, this being the most universal cause. The part is swollen and tender upon pressure, and if suppuration occurs the pain is severe, and cellulitis is apt to develop, involving the surrounding parts, or the joint may be involved, caries of the bones of the articulation resulting.

Treatment. If suppuration has not occurred, the part should be protected from pressure by a circular shield of felt or plaster; if suppuration has taken place, the part should be incised and drained, and if the joint is found diseased it should be curreted and dressed with an antiseptic dressing; if malposition of the toe exists, its position should be corrected by amputation of the head of the metatarsal.

Inflammation of Synovial and Serous Membranes. When the serous and synovial membranes are attacked by inflammation, the stage of congestion is accompanied by exudation of serum and fibrin from the surface, and the endothelial cells become swollen and detached in large numbers. The serous exudation may be sufficient to fill the entire cavity involved. There is a form of dry or fibrinous inflammation, without fluid exudate, in which the surface of the membrane loses its polish, becoming dry and red, and adhesions readily form wherever the surfaces are in contact.

In suppurative inflammation, pus is produced by emigration, and also by the detached endothelial cells. If fibrin is present, false membranes form on the surface and the membrane itself appears to be greatly thickened. At a later stage the proliferating cells invade these layers of fibrin and they become organized into connective tissue, and new vessels develop on them. Their tendency, however, is to disappear after a time, and the membrane returns to its original condition, unless the inflammation has been very intense, in which case the new connective tissue becomes permanent. Chronic inflammation of these membranes is marked by general thickening of all the layers, the formation of dense connective tissue in the fibrinous membranes, strong adhesions, and sometimes complete obliteration of the cavities, their endothelial lining disappearing entirely.

SYNOVITIS

Like other structures of the body the joints are subject to injury and disease and because of the nature and course of pathologic processes in them, one should bear in mind their anatomic construction.

The expanded ends of the bones in the joints are covered with a thin layer of cartilage and are bound to each other by a dense capsule which is firmly attached to the bones at their necks, where it is closely connected with the periosteum. The joint cavity is lined (excepting where additional fibrocartilages are present) with a synovial sac which sometimes communicates with a bursa.

Inflammations of varying intensity are of frequent occurrence; they maybe due to rheumatism or gout, to traumatism, to the action of microorganisms, or, to disturbances of innervation. They may be slight or severe, acute or chronic. They may terminate in resolution, in permanent new formations, more or less deforming and disabling, or in the destruction of the articulation.

Inflammations may arise in the joint structures proper or may extend to it from contiguous structures, such as the cancellous bone ends, the overlying tendons or the periarticular connective tissue. They may be largely confined to a single structure, the synovial membrane being ordinarily affected, or they may involve the whole joint.

Acute synovitis. Synovitis may occur as a result of a simple injury, such as a subcutaneous wound, a contusion, or a sprain. Exposure to cold and the presence of a movable cartilage are also common causes. Aseptic conditions in the synovial membrane seldom extend to the other joint structures (see "Arthritis") and heal with or without impairment of the joint, depending on the degree of inflammation.

Symptoms. The joint is painful, especially upon motion, and particularly so at night. It is swollen and tense and may be fluctuating. At the knee, the patella is floated up from the condyles and can be depressed upon slight pressure. The joint is held in a position of partial flexion which permits of the greatest ease, because of the diminished tension in this position.

Local heat and tenderness are not necessarily great, and constitutional symptoms, if present, are moderate in degree.

In the suppurative affections of joints, all of the above symptoms are intense and there is a general arthritis.

After a few hours or days the intensity of the symptoms subsides, the pain lessens, the swelling diminishes, as the effusion and extravasated blood are absorbed, the limb takes its natural position, and recovery promptly takes place. If there has been much hemorrhage into the joint, adhesions due to the organization of the clot may cause some restriction of motion.

Treatment. The joint must be placed at rest and an ice bag kept in constant contact. Even pressure with cotton and broad bandages often hastens absorption, but cannot at first be borne with comfort.

In rare instances aspiration of the effusion must be resorted to, but the certainty should exist that absorption is impossible, before a joint is punctured. The greatest care must be exercised in introducing a needle into a joint to avoid infection.

Chronic Synovitis. While it is true that an inflammation of a synovial membrane cannot long remain without extending to the other joint structures, the fact remains that symptoms peculiar to synovitis often persist for months. These are properly viewed as constituting a condition of chronicity. The active swelling and abundant effusion, belonging to the acute stage, subside, but an undue amount of fluid remains, with some pain and weakness.

If, with proper treatment and rest, these symptoms persist, there is an extension of the process to the bone ends and an exacerbation of symptoms.

The subsidence of a chronic synovitis generally leaves a weak and impaired joint, though pain may be absent. Movements, especially in extension, are restricted, and grating or cracking remain as evidences of the roughened membrane.

Treatment. The mere presence of a superabundance of fluid in a joint does not in itself constitute a diseased state, but may be the evidence of impaired circulation of the part. Absorption may occur with rest and tight bandaging, or with massage, friction, and baking, results may often be obtained. Certain cases resisting such procedures are best treated with a plaster of Paris cast to immobilize the part for several months. When the affection is of long standing and the joint is much distended it may be termed *hydrops articuli* or *hydrarthrosis*.

When, in spite of all the methods of treatment here described, the condition does not yield, very good results may be obtained by the aspiration of the fluid, and the injection of a few drams of a three per cent. or five per cent. solution of carbolic acid. This operation, though simple, requires every aseptic precaution, and should never be performed in the presence of any acute symptoms.

For other phases of Synovitis see Arthritis.

ARTHRITIS

The structures of a joint are: bone, cartilage, ligaments, synovial membrane and, in some cases, fibrocartilage. Hence, a joint inflammation is an inflammation of all of these structures, and is designated, *arthritis*.

The inflammation may begin in any one of these structures, but sooner or later, all are involved. The synovial membrane, however, when inflamed, seems to prove an exception to the rule in that inflammation may or may not extend from it to the rest of the joint. If such an extension does take place we have an arthritis.

We may therefore have two distinct classes of joint inflammation: (1) the varieties of synovitis, and (2) the varieties of arthritis. These inflammations may be acute or chronic.

In synovitis there is only the inflammation of the synovial membrane, while in arthritis there is inflammation of the synovial membrane plus inflammation of the bone covering (*periostitis*); of the bone (*osteitis*); of cartilage (*chondritis*); of bone marrow (*osteomyelitis*); and also a cellulitis of the ligaments attached to the joint involved.

Symptoms. The symptoms of arthritis are obviously more severe than those of a simple articular synovitis and are both local and general. The general symptoms arise from the absorption into the circulation of either bacteria or their toxins, and vary greatly in severity. There is either a toxemia or a septicemia, with the usual symptoms of a general sepsis.

The local symptoms are those common to synovitis and arthritis: pain, tenderness, swelling, heat, redness and loss of function. From these alone a differential diagnosis between synovitis and arthritis cannot be made. If, however, there is a sensation of crepitus conveyed to the examiner's hand upon passive motion, there is an arthritis present beyond doubt. This symptom is due to the destruction of the synovial covering of the bone ends involved, permitting contact of bone with bone. It is more common to chronic joint disease, but may also accompany acute conditions, especially if they are severe.

Symptoms peculiar to the variety of infection and the history as to duration, causation, course and number of joints involved, must be considered in making a diagnosis or prognosis.

Varieties. Besides simple traumatic arthritis, there are many constitutional disorders which affect the joints conspicuously; these are: tuberculosis, syphilis, gonorrhea, gout and rheumatism.

A prominent cause of many instances of arthritis heretofore regarded as rheumatic in origin, is now known to exist in any area of infection. Such "foci of infection" discharge a certain amount of infective material into the circulation, which may find lodgment in a joint and set up an acute process.

It has been proven in numerous cases that a so-called rheumatism will yield promptly to drainage of a chronic abscess, no matter how remote the location. Oral conditions especially have been found responsible for this form of arthritis. Abscesses at the apexes of teeth and pyorrhœa alveolaris, when properly operated, yield nothing short of miracles, in the way of relieved symptoms.

In addition to the varieties of arthritis already mentioned, those due to certain infectious diseases, such as measles, scarlet fever, typhoid fever, smallpox or erysipelas, should be included, as well as cases of neuropathic origin.

TRAUMATIC ARTHRITIS

Nonpenetrating and Penetrating

Nonpenetrating. Ordinary contusions or twisting at a joint, may result in the establishment of an inflammatory process within the joint, evidenced by much swelling and giving the sensation of fluctuation to the examining hands, indicating the presence of fluid within the synovial membrane. This occurs also when there is a detached fibrocartilage in the joint. The synovial membrane is thickened and there is an exudation of serum.

Sprains belong in this classification. These are simple, clean, inflammatory conditions.

Symptoms. These are generally limited to those enumerated as belonging to synovitis, except that the disability is more pronounced.

Treatment. Rest and wet dressings generally suffice to effect restitution in a few weeks.

Penetrating. Should the joint be injured by violence so that there is a loss of continuity of the tissues leading into the joint proper, there is every probability of infective material gaining entrance. These are serious accidents, though restoration of an efficient joint is possible, but when improperly treated or neglected, local destruction, or even loss of life may occur.

Penetrating wounds of joints usually occur in consequence of accidents with firearms, sharp tools, or falling upon sharp objects. Frequently, penetration of a joint follows suppuration in the immediate neighborhood.

Symptoms. The extent of the injury, the particular joint involved, and the nature of the vulnerating body will affect the train of symptoms. An escape of synovial fluid, pain and some swelling will occur even with a very small penetration. Should the joint escape infection, the synovitis quickly subsides and recovery takes place with little or no impairment of the functional value of the part. The opening in the capsule closes, the extravasated blood is absorbed and the synovial surface is again smooth. If, however, the wound has been inflicted with an unclean instrument, or if at any time before healing it becomes septic, a very different and graver condition obtains.

Septic Arthritis. Infection with bacteria of suppuration, chiefly the staphylococcus albus or the streptococcus pyogenes, produces an acute arthritis which frequently, despite the most careful treatment, will result in the destruction of the joint, and not seldom in the loss of life.

The infection may occur in one of several ways: (1) directly through a dirty instrument, or the lodgment of infective material in the tract leading to the joint cavity; (2) by the extension of a suppurative process, either of the bones or soft tissue adjacent; or, by (3) the deposition into the joint of infective organisms circulating in the blood stream.

Symptoms. However produced, large numbers of organisms are present and a high grade of inflammation ensues. An abundant amount of pus is soon formed; the synovial membrane, the bone ends and the joint capsule are actively inflamed, and soon become disorganized. Perforation of the capsule is followed by infection and suppuration of the tendons and other structures about the joint, which soon affects the superficial structures and forms an opening through the skin. The pain is intense, generally worse at night; the swelling is great and fluctuation is distinct; the skin is red and hot, and the parts above and below are edematous. Any attempt at motion increases the suffering.

With these local symptoms there is an accompanying train of constitutional symptoms which may eventuate fatally. At first there is a chill, or a sensation of chilliness after which the temperature quickly runs up several degrees, and either remains so, or goes down and up several times in twenty-four hours, as in other septic conditions. The pulse may be strong and full at first, but soon becomes rapid and weak. In very acute cases, death from septicemia may occur in a few days.

In ordinary cases, drainage of the pus, either naturally or artificially, will result in a remission of the symptoms both locally and generally.

Treatment. In this, as in other suppurative processes, safety lies in the prompt opening of the abscesses and the evacuation of the pus, thus accomplishing free drainage, with subsequent disinfection by means of applications or irrigations. Immobilization of the parts and rigid antisepsis will generally yield good results as to life, though recovery with ankylosis is the rule. In the most severe cases, constitutional symptoms are so grave as to warrant immediate amputation above the infected joint.

Tubercular Arthritis. The great majority of chronic joint diseases are tubercular in origin, the tubercle bacilli being deposited in any of the joint structures, or in structures contiguous to a joint; with children, very frequently in the bone substance.

Whether the tubercular process originates in the joint cavity itself or outside of it in the surrounding tissues, destruction of the articular ends of the bones is usual.

The parts become thickened and edematous; there is a gelatinous or cheesy appearance, in which the membrane, cartilaginous bone ends, capsule, and ligamentous structures all share. Frequently the synovial membrane is studded with miliary tubercles and its cavity is filled with an abundant serous secretion. The contour of the joint becomes globular or spindle shaped, because of the atrophy of the parts above and below it and the swelling of the periarticular structures. The skin becomes white and thick because of the obliteration of the superficial vessels and because of its edematous infiltration.

Symptoms. Pain is, as a rule, but slight in the strictly synovial stage of tubercular arthritis, but when the bones are involved, it is severe, though acute symptoms, such as heat and redness, are lacking.

Deformity is a constant accompaniment of the disease; its degree is greater or less according to the joint affected, the extent of the disease, and the treatment pursued. It is due to the natural tendency to assume the position of greatest ease; to the softening and destruction of the ligaments, and to the effort on the part of nature to immobilize an injured member by means of tonic contraction of the muscles. These causes often result in the creation and persistence of a malformation and malposition of the part.

Cheesy degeneration and liquefaction take place in more or less degree, and though their occurrence is often not evidenced by any aggravation of the symptoms, sinus formation with persisting discharge occurs.

When these sinuses occur, they generally become infected with other pus producing organisms, and aggravate the condition considerably. In the course of months or years, many such openings may occur through which masses of soft tissue or bone, either carious or necrosed (*sequestra*), may be discharged.

Diagnosis. This may be easy, difficult, or impossible, depending on the duration, the joint involved, and the character of the disease in any individual case.

At times it is impossible to differentiate from syphilis, which, however, is quite uncommon, but with which tuberculosis has many symptoms in common. The history of the individual, and a blood examination will generally suffice. If the disease is advanced to the stage of abscess and sinus formation, there can be no doubt as the nature of the trouble.

Very often the disease in the articular ends of the bones advances slowly, giving very little pain and no appreciable swelling or atrophy. There may be only an unwillingness to use the part very much, and the disease may very well be overlooked. In such insidious cases a diagnosis can be reached by aspiration and subsequent examination of the serous fluid for tubercle bacilli. An X-ray will show the rarifaction of the bony structures and the thickened periosteum.

The course of tubercular joint disease is entirely dependent upon its extent at the time it is recognized, and the treatment pursued. It is of paramount importance that attention be given any persisting pain or discomfort in or near a joint, and that rest and every diagnostic aid be employed before pronouncing a case hysteria, neuralgia or "growing pains." In a few cases the process can be arrested and little or no diminution of function remains. This, however, is the exception; there is usually destruction of the intra-articular cartilages, and of the synovial membrane, and the formation of bands of great density, which impair the motion of the part even to rigidity (*fibrous ankylosis*). The restriction of motion may be absolute if ossification of the granulation tissue lying between the epiphyses unites their eroded ends (*bony ankylosis*).

At times, though recovery seems to have been secured, a sinus may persist because of some slight area of remaining caries, or because the tract itself is tubercular. In other instances a recurrence may follow after months or years of quiescence. This may be due to the setting free of encapsulated organisms, or because of a new infection at a point of least resistance.

Treatment is that of tubercular disease in general. The most essential features in the conduct of these cases are rest and the establishment of ideal hygienic conditions. Forced feeding, sunlight and air, play as important a part here as in pulmonary tuberculosis. Absolute rest of the part can be secured only with the aid of plaster of Paris braces, or splints of other materials. Such immobilization should include the joints immediately above and below the one affected. Hyperemia, by the use of a rubber bandage above the joint, or by baking of the joint, is of great value.

In the majority of instances these methods will yield good results in from six months to a year. Operative interference will be necessary in addition to the above, where caseation and secondary infection have occurred. Thorough drainage of the infected joint, either by widening already existing sinuses, or by free incision followed by irrigation, will frequently be necessary.

Joints Generally Involved are the larger ones of the extremities, but this does not preclude the possibility of any joint being the seat of a tubercular inflammation. The vertebral articulations and the digital articulations of the feet and hands are commonly affected. In children, the hip joint is the one most attacked; frequently the knee, ankle and elbow are affected in the order given.

In nearly all cases of arthritis of tubercular origin the original focus of infection is located in the bone, though the synovial membrane, or an adjacent osteomyelitis, may be the first point attacked.

Syphilitic Arthritis. This is rather a rare condition, but must be differentiated from tuberculosis, because of its slow onset and progress, and because of the mildness of the symptoms and the spindlelike shape of the joint. There is usually but one joint involved and eventually a dark fluid will escape should sinus formation occur.

Diagnosis will generally be known in advance from the history, through a Wassermann test of the blood, or an X-ray picture will often be of value.

In syphilis, the original focus of infection in a joint will be found in the soft tissues, while in tuberculosis, the articular ends of the bone are first involved. An examination of the discharged fragments of tissue in syphilis will show a round cell infiltration; in tuberculosis, possibly typical tubercle tissue.

Treatment by anti-syphilitic remedies, if successful, will also indicate the nature of an obscure case, a pronounced response to such treatment being a positive diagnostic aid.

Gonorrhoeal Arthritis. This affection is nearly always very acute, beginning as an acute synovitis and extending to the articular fibrocartilages at an early date.

Constitutional symptoms nearly always accompany this variety of arthritis, a chill and high temperature being the rule.

This condition is often called gonorrhoeal rheumatism. It is due to the lodgment of the gonococcus of Neisser in the joint, from the blood stream.

Gonorrhoeal arthritis is a form of septic arthritis, its pathology and symptomatology being in many respects the same. It may, in favorable cases, limit itself to the synovial membrane, in which event the symptoms will yield more readily to treatment, though the affection in any event is an acute one, and a diagnosis as to extent is difficult to make owing to the extreme pain of even slight motion.

Symptoms. These are similar to those of septic arthritis, except that usually only one joint is affected and the existence of a gonorrhoeal infection can always be determined. Both knees, or both ankles, but more commonly, only one joint, are affected, accompanied by severe constitutional symptoms. There rarely occurs any indication of sinus formation or of spontaneous drainage in this variety of arthritis, and it is held by many, that in cases where this tendency exists, there is a mixed infection, other pus producing organisms being present.

Treatment. The original infection of the urologic tract must receive the utmost care, in order to eradicate the supply of germs to the circulation. The injection of anti-gonococcic sera or vaccines finds its best application in these cases. The local treatment consists of rest and immobilization of the extremity affected.

The application of either extreme heat or cold to the joint is agreeable and efficacious.

There are many reasons in justification of either of these treatments over the other, but in general it may be said that, in the acute stage, cold is better, while in the latter stages, heat will accomplish more to establish easy motion of the part and to lessen the danger of ankylosis.

Active or arterial hyperemia by baking, is especially valuable in the subacute stage.

Prognosis. In those cases in which the pain and swelling is severe and the constitutional symptoms alarming, we may always expect a true arthritis to exist. In these cases much exudate is formed in the joint, which upon organization, leads to fibrous bands and limitation of the joint function (*fibrous ankylosis*).

In the milder cases, ankylosis is the exception, if proper remedial measures are carried out.

Rheumatic Arthritis. Rheumatic articular affections are common, and are both acute or chronic. In the light of recent investigations it is believed that many of these cases are due to foci of infection in various parts of the body which pollute the blood stream with organisms which subsequently find lodgment in either the organs or joints. Infections existing in the tonsils and teeth roots have been shown to act in this way. There may, however, be cases directly attributable to rheumatism, though these are not so well understood.

Acute Rheumatism. One or several joints may be attacked simultaneously. Subsidence of the inflammation may occur, while others are becoming inflamed.

Symptoms are those of acute synovitis; suppuration never occurs unless there has been a mixed infection, and limitation of motion is a rare sequela. The pain, swelling and tenderness is extreme, and the constitutional symptoms, while being severe are not usually grave. In the *chronic variety*, on the other hand, there may be limitation of motion due to the formation of bands and adhesions after months or years of inflammation. This variety may start as such or may begin as an acute condition.

Treatment. The treatment, besides local rest and heat, consists of the administration of antirheumatic remedies and hygienic precautions.

Diagnosis will rest largely on the blood examination for circulating organisms, the general examination for foci of infection, and the family history.

Gouty Arthritis. Whatever may be the essential nature of gout, its manifestations are common in the smaller joints, such as the fingers and the metatarsophalanges of the great toe. Deposits of urates, chiefly sodium urate, take place in the connective tissue of the joint and also in the cartilage. Consequent upon the irritation of these salts, there is an increase in the connective tissue followed by contraction, impairment of motion, and alteration in the shape of the joint. Repeated attacks of acute inflammation occur, of greater or lesser intensity, and the uratic deposits attain a considerable size, occasionally forming abscesses or ulcerations in the overlying skin.

Like rheumatism, gout is a manifestation of a constitutional state, and requires medical care.

Infective Arthritis. These are the arthritic manifestations of diseases as smallpox, scarlet fever, typhoid fever, measles and erysipelas. They are due to infective material deposited from the circulation, and are in every way similar to septic arthritis, which see. There are always suppurative synovitis and osteomyelitis, with a consequent ankylosis of bony structure. The constitutional symptoms are very intense, and free incision and drainage is indicated.

Neuropathic Arthritis. (*Charcot's Disease*). This is a peculiar osteoarthritis observed in patients with locomotor ataxia. The disease is an acute one, so far as objective conditions are concerned, there being no pain or constitutional derangements of consequence. Without any injury having been received, the joint, particularly the knee, suddenly swells, the intra-articular effusion becoming abundant. This may soon be absorbed and with it the articular ends of the bones wear away and break down into small fragments. The limb becomes atrophied and shrunken, and the joint itself becomes weak, often flail.

This disease seems to be due to nutritive changes in consequence of changes in the spinal cord nerve centres. There is no satisfactory treatment and the patients must be kept in bed.

CHAPTER X

DISEASES OF THE BONES

Congenital Defects of Bones. Various congenital deformities of the limbs occur because of interference in various ways with the proper and normal formation of these cartilaginous masses. If, for any reason, the cause of which in most cases is not clear, any of these cartilaginous masses fail to be formed in the embryonic tissues, naturally no ossification can occur, and in such cases there may be a partial or complete lack of development of the corresponding bone. The amount of this congenital deformity may vary from the absence of an entire foot, to the absence of one or several digits, or one or more phalanges.

The deformities produced by such a failure to deposit the cartilaginous base of the bones are very numerous, and in some cases lead to great deformity and loss of function. This lack or increase of the reformation in cartilage, results in most extraordinary deformities.

No special type of deformity merits special attention; the condition in each case must be decided by inspection and X-ray examination.

In many of these cases, especially where the lesion affects the digits, the capability of the individual is but little impaired, whilst in other cases, where bones are absent, marked deformity and impairment of function may occur. Some of the cases, notably webbed toes, are comparatively easily corrected; other cases however, offer little chance of sufficient cosmetic or functional gain to make a surgical operation necessary or desirable.

Atrophy of Bone. Various causes may lead to atrophy of bone. The method by which atrophy is brought about is peculiar, and is due to the action of special giant cells, called osteoclasts. Wherever extensive atrophy of bone takes place, microscopic inspection shows such giant cells lying closely adjacent to the trabeculae of the bone which is being resorbed, and the trabeculae in that immediate vicinity slowly disappear under the action of these giant cells. Their action is very similar to the action of giant cells in the soft tissues about absorbable foreign bodies. This process is called *lacunar resorption.*

In old people the amount of absorption oftentimes is very great; the process is then termed *senile atrophy*. It may be marked in the skull and in the long bones, and in many cases of fracture of the neck of the femur, a moderate amount of lacunar resorption precedes the fracture which results from slight violence. In certain cases this resorptive process in old people is extreme, and leads to great fragility of the bones, with repeated fractures from slight violence, which under ordinary circumstances, would cause no injury at all.

A mere lack of use of bones may also lead to a certain amount of atrophy from lacunar resorption. This may be seen after amputations, where the stump of bone which is left from the amputation slowly undergoes lacunar resorption and sometimes a marked diminution in size. The same thing may also be seen in the bones of people who for long periods of time have been deprived of the use of their limbs, either by the application of apparatus around fractures, or by disuse for other reasons.

Lacunar resorption also occasionally follows lesions of the central nervous system, part of the atrophy being due to disuse of the limbs from the paralysis, and part of it also being dependent in some indirect way upon the nerve lesion.

Atrophy of bone also may be brought about by pressure. It is to be remembered that the bone, as a matter of fact, is not a perfectly rigid material, but that processes of new formation and resorption are constantly taking place, even under normal conditions. If, for any reason, bone is put under constant pressure, a certain amount of readjustment of the bony constituents takes place in order to adapt the bone to its altered condition. The most striking example of this sort of atrophy is perhaps the Chinese ladies' feet, where the bones, being bent into an abnormal position, beginning early in childhood, ultimately show enormous deformity and an entire rearrangement of the trabeculae of the bone. The same thing also may be seen occasionally after pressure and deformity from contracture of muscles or from the pressure of scars. This process, which ordinarily leads to loss of function, in a certain limited number of cases aids function, for whilst certain fractures of the joints may lend to deformity of the articular facets of those joints, by absorption of certain portions and new formation in others, a readjustment of the joint surface may take place, so that a marked increase of function may occur.

A certain amount of atrophy also may be brought about by the pressure and development of tumors.

Hypertrophy of Bone. In many cases new growth of bony tissue is due to the new formation of periosteal bone, and is an expression of an attempt at repair of one or the other of the numerous destructive processes. In other cases true hypertrophy of the bone, with no connection with any reparative process, may occur.

A notable example of this is seen in the growth of bone which sometimes occurs after amputation, especially in young people. The increased size of the bones which is seen in many definite diseases will be mentioned under the proper headings.

Caries and Necrosis. Various pathologic processes produce destruction of bone. The destructive process may cause the death of large areas of the affected bone at once, and in that case, a large fragment of necrotic bone may remain in situ and still maintain its contour. Destruction of bone of this sort is described by the clinical term *necrosis*.

Other processes cause a gradual molecular softening and destruction of bone, which ultimately may be very extensive, but at no time is there present any appreciable large mass of bone. Destruction of this sort is described by the clinical term *caries*.

As a means of differentiating clinical conditions, the use of these two words is desirable. As a clinical term, *necrosis* usually means destruction by pyogenic infection, and *caries*, destruction by the gradual extension of a tuberculous process. This clinical distinction, however, is not an exact one, because destruction of large areas of bone, described as necrosis, is occasionally brought about by syphilitic infection, and rarely by tuberculosis, whilst molecular destruction of the bone is brought about by a considerable variety of processes, the chief of which, it is true, is tuberculous infection, but actinomycosis and syphilis may both lead to the gradual disintegration of the bone, without the formation of large necrotic masses of bone.

The presence of necrotic bone connected with the surface of sinuses, from which comes a discharge of pus, should always lead to the consideration of tuberculosis, actinomycosis, and syphilis. The presence of large sequestra of bone should immediately suggest the presence of osteomyelitis or of syphilis.

Treatment. The details of the treatment of the various forms of destructive processes in bone will be found under their special headings, chiefly under osteomyelitis and tuberculosis.

In all cases of caries it is desirable to remove completely the softened areas in the bone. This may be done by curettment and drainage, or by excision of the entire bone, or series of bones, in certain cases, or rarely by amputation.

The difficulty in all these cases is to recognize the exact limits of the carious process. It must be borne in mind that at the time of operation upon carious bones the field of vision of the surgeon is almost always limited; moreover, the bleeding which always takes place from the bone-marrow in such cases, also obscures the field, and even if these two causes were not present, it is frequently extremely difficult, by naked-eye examination to determine the exact limits of the destructive process. As a general rule, it can be said that the carious area is at least a quarter of an inch wider than appears upon visual inspection.

In cases of necrosis with large bone defects, the difficult thing is to cause a growth of the bone toward the central cavity after removal of the sequestrum. The various methods applicable to such cavities are mentioned in detail under "Osteomyelitis."

PERIOSTITIS

Acute Periostitis. The older text books always laid great stress upon the occurrence of an acute infectious inflammation of the periosteum. Acute suppurating periostitis alone does not occur, and most of the cases which have been described as such are really mild cases of superficial osteomyelitis, with abscess formation beneath the periosteum, and possibly slight inflammation of the periosteum itself.

These cases ordinarily lead to only a slight destruction of the outer layer of the cortical bone.

Symptoms. These are the same as in acute osteomyelitis, except in a very much milder form. There is usually a rise of temperature, oftentimes with a chill, with circumscribed tenderness over some portion of the shaft of one of the long bones.

Treatment. Incision over such an area shows an elevated periosteum, with a small, localized abscess beneath it, with bare, white, somewhat vascular bone cortex. Incision alone in most cases suffices to cure the disease, although if the process has extended sufficiently deep to cause a superficial necrosis of the outer layer of the cortex, removal of a small sliver of necrotic bone may be necessary.

Chronic Periostitis. A long-continued and chronic irritation of the periosteum, sufficient to cause a proliferation of the osteogenetic cells of the periosteum, is common in a great many diseases. A chronic thickening of the periosteum with a new formation of bone, is seen frequently after traumatism, blows or contusions; sometimes after the occurrence of superficial abscess of the soft tissues in the immediate vicinity of the shaft of the long bone, described as chronic ulcer of the surface of the tibia; or after certain infectious diseases, notably syphilis. It also may occur after various other local infections. In such cases the thickening of the periosteum ordinarily is pretty sharply localized.

A general thickening over the periosteum, and over several or many of the bones of the body, also occurs in the disease known as *toxic osteoperiostitis ossificans,* seen in diseases with long continued suppuration. It also is common after syphilitic disease, either congenital or acquired.

Symptoms. The symptoms of chronic periostitis with new formation of bone are invariable. In a certain number of cases there is a constant, heavy, dull pain, at the point of thickening, with at times more or less acute exacerbation; at other times the lesion is associated with no pain whatever, and the patient's attention is first called to the disease by the presence of the enlargement of bone. Recognition of the condition may depend upon X-ray examinations for indefinite pains in or over the bone.

Chronic periostitis is not really a disease itself, but a manifestation of the reaction of the periosteum to some irritant.

Treatment of the condition depends, first of all, upon a recognition of the cause and a removal of the cause, when possible. In many cases, especially those in which no pain is present, nothing in the way of therapeutic measures can be done.

The chronic thickening of the periosteum, seen in many definite bone diseases, will be mentioned under those diseases.

Osteomyelitis. Infectious osteomyelitis is acute suppuration of the bone, always due to the infection of the bone marrow by pyogenic microorganisms. The process is essentially like the process seen in furuncle, and begins in the marrow of the alveolar spaces, which communicate freely with each other, but are enclosed by a dense shell of cortical bone. Hence the process may quickly at first involve the entire marrow of an infected bone, because the products of bacterial infection are retained in this dense shell, while the primary focus can only be reached by extensive bone operation.

Most cases are due to the staphylococcus pyogenes aureus and a few to the streptococcus. Typhoid bacilli may cause suppuration. The infecting organism is present in pure culture but sometimes a mixed infection occurs, and such cases are said to be severe.

In cases of chronic osteomyelitis with open sinuses and exposed bone, a great variety of organisms, pathologic and saprophytic, may be present. Hence infectious osteomyelitis is not a specific disease, but is acute inflammation of bone that may be produced by any one of a variety of pathogenic organisms, or by a mixed infection.

Any pyogenic organism which can be carried in the blood may be deposited in the bone and produce suppuration. Some of these organisms may settle by preference in the bone marrow, others beneath the periosteum, or in the joint.

Certain general causes favor the occurrence of osteomyelitis. Children are chiefly affected and it occurs in boys about three times as often as in girls. Acute osteomyelitis frequently occurs after injuries of moderate severity, because such injuries may lower resistance of the bones and make them unusually susceptible to pyogenic infection. One of the commonest causes is the infection of a compound fracture, and before the days of asepsis, such cases were very frequently fatal. Under modern methods the infection, when it does occur, is generally slight, although the destruction of bone may greatly delay healing and may lead to the formation of small sequestra and indurating sinuses. Infection of a similar sort may occur subsequent to amputation.

Osteomyelitis nearly always begins in the diaphysis of the long bones, usually near the epiphyseal line. This is an important point, clinically, because tuberculosis practically begins in the epiphysis. In rare cases, however, osteomyelitis begins in the epiphysis, and so may simulate tuberculosis. The femur and tibia are the bones most frequently attacked, but no bone is exempt. Usually only one bone is affected, but cases of multiple bone infections are not rare.

The primary area of infection is always in the bone marrow. The bony trabeculae and the cortex are destroyed only secondarily. The process nearly always begins in the diaphysis, but then may extend into the epiphysis and produce suppuration of the joint. Once the organisms have gained access to the marrow, they produce a toxin which causes necrosis of the adjacent marrow cells, and this necrosis may extend over a very considerable portion of the bone before marked infiltration with leucocytes occurs. The infection usually extends quite early through the dense cortex by way of the Haversian canals, and produces an inflammatory exudation and suppuration between the periosteum and the outer layer of the cortex, which is designated*subperiosteal abscess*.

Such an abscess may strip the periosteum from the bone over very extensive areas. The infection may then extend to the adjacent soft parts, muscles and subcutaneous tissue, and form an abscess outside the periosteum.

If, from spontaneous opening of the abscess or from operation, a fatal result is avoided, the infective process may be limited and the process of repair may begin.

As a rule, a portion of the infected marrow and cortex become completely necrotic, and the lime-bearing portion of the bone persists as a more or less extensive sequestrum.

The periosteum in the early stages may be separated from the bone by a collection of pus, and in such cases it appears as a thin fibrous membrane beneath the muscles, separated from the bone by the abscess cavity.

Secondary changes occur in the soft tissues surrounding the seat of an acute suppuration of bone. During the acute stage there may be a definite abscess of the soft parts, with an infiltration which simulates phlegmonous inflammation, or, by rupture of the abscess, various sinuses may be formed leading down to the necrotic foreign body. In long continued cases the skin and subcutaneous tissues become thickened by the formation of scar tissue, due to the presence of the involucrum and the persistence of sinuses, and by thickening of the soft tissues, an affected limb may for years be nearly twice its normal size.

Symptoms. The disease usually begins with a sharp onset, the first symptom being a sudden localized pain in the vicinity of the epiphyseal line, or in the shaft of some one of the long bones. This pain is extremely intense, and in typical cases is most excruciating.

Motion of the joints at this time is not painful, but the pain produced by percussing the bone, even lightly, may be intense. An extremely valuable diagnostic point is continued gentle pressure at some point over the shaft of the bone at a distance from the point of greatest constant pain.

Usually, at a very early period, there appears swelling of the soft parts about the bone. This swelling, at first, is neither hot nor red, but soon becomes edematous, red, and shows pitting on pressure, and at that time may simulate acute phlegmon.

In some cases the adjacent joint early becomes tender, hot and swollen, and this may occur even when there is no real extension of the infectious process to the joint itself. If extension does occur to the joint, swelling, tenderness, and pain on motion become more intense. The temperature usually is elevated to a considerable degree—103°F. or 104°F.—and usually the pulse is greatly accelerated. Evidence of constitutional disturbance and absorption of infectious material occur early. The tongue is dry, coated and tremulous; the face is drawn and flushed. Delirium of a mild type is a very common symptom, and in some cases this delirium may persist for a considerable length of time after the bone has been drained. Abscess of the soft parts may give deep or superficial fluctuation. Sinuses may appear. The leucocyte count is usually very high—25000 to 35000,and chiefly of a polynuclear type.

Such a clinical picture is perfectly distinct, and it is difficult to overlook typical cases, especially after the fluctuation in the soft parts has occurred. The diagnosis of early cases, however, is sometimes very difficult, and even in the hands of experienced men, who have the lesion in mind, is frequently impossible. Even in severe cases, occasionally the pain itself is not severe for several days, when there may come a sudden exacerbation of symptoms.

In the chronic stages of osteomyelitis the symptoms are usually characteristic. The limb is enlarged, the enlargement being partly due to thickening of the soft tissues, but chiefly to the formation of the involucrum. Usually running down to the sequestrum, are enormous sinuses, from which comes a foul, purulent discharge. On passing a probe, dead bone can be felt at the bottom of the sinuses. It must be borne in mind, however, that in a great many cases, after attacks of osteomyelitis of moderate severity, small localized abscesses are formed in the shaft of the long bones, with no sinus communicating with the surface. An abscess of this description, as has already been stated, is always surrounded by a wall of dense bonelike cortical bone.

Such an abscess may persist for years with no symptoms beyond a moderate enlargement of the shaft of the bone at the point of abscess, and the enlargement may be so slight that it is not recognized by the patient. In other cases the entire shaft may be enlarged, but the bone may not be tender. In most cases, however, such a localized abscess sooner or later gives rise to recurring attacks of pain, which, as a rule, are extremely violent. The intervals between such attacks may vary from days to weeks, or to months, or even to years. The attacks of pain may come on, apparently, perfectly spontaneously. Associated with these attacks of pain, the bone over the abscess usually is exceedingly tender to touch. With the attacks of pain may come a rise of temperature, or in some cases, there may be no disturbance of the general condition. This kind of abscess may be of small size, no larger than a pea, or may involve a great portion of the shaft of the bone; in such abscesses no definite sequestrum may ever form.

The recognition of such conditions depends upon recurrent attacks of violent pain over circumscribed areas of bone, with or without constitutional disturbance, and nearly always with extreme local tenderness.

Treatment. In the acute stage there is suppuration of the marrow, more or less extended throughout the shaft, with often a subperiosteal abscess and perhaps abscess of the soft parts.

The indications are the same as in any other acute suppuration; the pus must be evacuated and the bone cavity drained. This demands not only an incision into the soft parts, but an opening into the shaft of the bone. If a piece of necrotic bone is present, it should be removed.

In the chronic stage there is usually an old necrotic shaft perforated by sinuses, and often freely movable, inclosed by a shell of dense periosteal bone. The sequestrum must be removed, but the bony defect fails to heal, and for months persists as a filthy, discharging cavity, with the constant danger of secondary infection and phlegmon, or erysipelatous inflammation. The healing of this cavity is very difficult and requires a very long time.

Many methods have been tried for the filling of these bone cavities with blood clot, iodoform and oil of sesame, but they have not been successful, because it is almost impossible to render such cavities absolutely aseptic.

Tuberculosis of Bone. Tuberculosis of bone is always dependent upon infection of the marrow of bone by the tubercule bacillus. This germ obtains entrance to the bone marrow and causes the formation of miliary tubercules which arise from the proliferation of the connective tissue of the marrow around the primary tubercule. Other secondary tubercules are formed by extension of the tubercule bacillus. The centres of these tubercules become caseous, and, by fusion of adjacent caseous areas, also cause softening in the bone marrow.

The tuberculous process, as a rule, begins in the epiphysis in the long bones, and may affect any of the bones.

Symptoms. In cases of tuberculous disease confined to the bones alone, the first symptom usually is pain, which ordinarily is not severe and has a gradual onset. Oftentimes, at first on palpation, no difference in the shape of the bone can be detected.

Toes affected by a tuberculous process, slowly enlarge at first without heat or pain; ultimately the skin becomes thickened, and reddened, and the digit is painful to pressure or motion. Oftentimes the skin is perforated at one or more points by sinuses lined with tuberculous granulations, through which caseous pus is discharged.

The diagnosis in these cases always lies between tuberculosis, actinomycosis, syphilis, and osteomyelitis, and exact determination of the origin of the cause oftentimes can be made only by inoculating animals with a discharge from the sinus, or by detection of pyogenic organisms, or of the miliary tubercule, the histologic unit of tuberculosis, or by detecting the peculiar yellow bodies seen in actinomycosis.

Treatment. From a clinical point of view tuberculosis of bone should be considered in the same category as malignant disease, and the indications for treatment in all cases of tuberculous bone disease are the same as in malignant disease; which is, complete removal of the infected area, whenever it is possible.

In some cases the mere opening and curetting of tuberculous areas in bone is oftentimes enough to set up sufficient reaction in the bone and in the surrounding tissues, to put an end to the tuberculous process. Complete resection of bones may at times be avoided by this treatment.

In addition to the local treatment of opening, curetting and drainage, or the complete excision of the bone, the greatest care should be employed in the management of the general hygiene of the patient, including feeding and fresh air. Often removal to a climate which is unfavorable to the development of tuberculosis in general, is also extremely desirable.

Syphilis of Bone. The lesions produced in bones by syphilitic infection may be congenital or acquired, and, as in other syphilitic lesions, the manifestations may be protean.

Most children with congenital syphilis, show an irregularity of the epiphyseal line, which results in the latter becoming markedly toothed, instead of constituting a straight line across the bone, at right angles to the long axis of the shaft.

Besides the irregularity of the epiphyseal line, three other changes are seen in the bones of syphilitic infection. The most common lesion is one which affects the periosteum and leads to the formation of periosteal bone. This periosteal formation may occur either in congenital or in acquired syphilis, and it may affect one or many bones. In some cases there is an enormous thickening of the epiphysis of the bones, and as a result of the epiphyseal thickening, secondary changes in the joints occur, so that the thickening of bones and the changes in the facets of the joints, suggest fracture or dislocation. In other cases, the thickening affects only the shafts of the long bones, generally of the leg or arm, although no bones are exempt. In some cases, both in the congenital and acquired forms, there may be marked proliferation of the endosteum of the bone, with or without thickening of the periosteum, although thickening of the periosteum usually is present. This process, as a rule, affects one bone in its entirety, and most commonly affects the bones of the lower leg, notably the tibia. As a result of these changes the bones are enlarged and thickened, and in some cases, from endosteal thickening, the marrow canal is very largely or entirely obliterated. In some cases true gummata of the bone are formed. These gummata may appear in the spongy portion of the bone, sometimes in the shaft, or in the epiphysis. They also appear to be formed in the lower layers of the periosteum and lead to circumscribed nodular thickenings on the surface of the bone.

Symptoms. These vary with the different pathologic conditions present. The periosteal thickening may occur at any time of life over any bone of the body.

The presence of circumscribed periosteal thickening of bone in itself should always lead to the suspicion of the presence of syphilis.

Pain, as a rule, is only very slight, and the diagnosis depends upon the history and the detection of other syphilitic lesions.

The cases in which there is both endosteal and periosteal thickening, occur chiefly in children and are of a congenital nature.

The physical symptoms are very characteristic. The bone usually affected is the tibia, which is enlarged to a most marked degree, and often shows a pronounced bowing forward, similar to the bowing and thickening of the tibia seen in osteitis deformans. The bone is extremely dense and obviously heavier than normal. The bones are moderately tender to pressure, but have nothing like the extreme tenderness noted on pressure in osteomyelitic bones.

In cases of gummata of bones the symptoms vary. In some cases the gummata are on the surface of the bone, especially the sternum, and at times on the long bones. In such cases there appear a softening and reddening of the skin about the affected area, which remains indolent for a long time.

If such an area opens spontaneously, or is opened by incision, the contents are seen to be composed of a yellow, rather gelatinous material, quite like the caseous material from a tuberculous abscess.

Treatment. In most cases the regular anti-syphilitic treatment is indicated. In cases of periosteal thickening, the results vary with the time at which the treatment is begun. In the early cases, a thorough anti-syphilitic treatment may lead, after a varying length of time, to complete disappearance of the newly formed periosteal bone. On the other hand, if the periosteal process has lasted for a long time and the bone has become densely cortical, although anti-syphilitic treatment may lead to a diminution of the localized pain, the dense bone does not disappear. In cases of combined endosteal and periosteal thickening, the pain usually disappears under anti-syphilitic treatment but the changes in the bone persist.

Osteomalacia is an acquired disease which causes marked softening and changes in the bones. The disease begins irregularly and often progresses with or without remissions. The progress is more marked during pregnancy. The first sign is pain in the bones, which is increased by pressure, and this is especially true of pressure over the ribs. There are also muscular cramps and contractures.

Osteitis Deformans. (*Paget's Disease*). This is a chronic disease of the bones and may affect one or more bones of the body. The onset is insidious, and before actual deformity occurs, long indefinite pains in the legs may have existed, with occasional tender points over the bone.

The bony changes are first noticed in the bones of the legs and are most marked in the tibia, femur and fibula. As a result of structural changes, these bones become bowed, while their internal trabecular structure is altered.

The extent of the affection in the bones of the legs varies a great deal and usually is not symmetric. The lower extremities are bowed outward, and also are usually bent forward, the curves being due to changes in the femur and the lower leg.

Treatment. In the absence of any knowledge as to the cause of the disease, the treatment of osteitis deformans must be largely symptomatic. Certain drugs have been recommended; among these are iodide of potash and arsenic. Most such patients are in poor general condition, and effective feeding often gives marked relief of the symptoms from which they are suffering.

For severe pain, counterirritants are valuable, especially the actual cautery. Massage is of use in some cases for improving the general condition.

Tumors of Bone. All the primary tumors of bone are of the connective tissue group, but various secondary tumors of epithelial origin may occur.

Osseous tumors may arise from the periosteum or from the marrow. If they arise from the periosteum they may extend early to the adjacent soft tissues and involve and destroy them. If the tumor arises in the marrow, it is for a long while cut off from the adjacent soft tissues by the thick cortex, and about the extending medullary tumor may also come a reactive proliferation by the periosteum, so that as the tumor extends it still may, for a long time, be surrounded by a shell of bone which prevents infection of the soft parts. After a time, however, the reactive periosteum shell usually becomes perforated at one or more points, and then the medullary tumor extends to the adjacent tissues. The cause of these tumors is absolutely unknown.

Fibromata are not very common tumors of bone. They arise generally from the periosteum and are most common about the face, and are rarely seen in the long bones. Many of these tumors are closely allied to some of the fibrous forms of sarcoma, and it is often difficult to distinguish them histologically.

Chondromata are fairly common tumors of bone. They may appear externally to the cortex, or sometimes they grow in the medullary canal. They may arise directly from the marrow, probably from remnants of the provisional cartilage cells. They also appear frequently to arise from the epiphyseal line.

Chondromata appear generally as multiple masses, nodular in shape, and are frequently seen on the lower leg, about the knee joint. They usually are painless, firm and hard, and not tender to pressure.

Treatment consists in removal by operation.

Osteomata are bony tumors which generally arise by growth of the periosteum, and form solid bony masses external to the cortex of the bone, when they are called *exostoses*.

The density of the bone composing the tumor varies a great deal, some being very hard and ivorylike, while others are like the cellular marrow of the long bones.

Osteomata may be surrounded by a layer of fibrous periosteum or, in certain cases, beneath the periosteum appears a layer of cartilage producing the so-called *exostosis cartilaginea*. The latter formation is the one which is most common in the vicinity of the epiphyseal line of the long bones, notably of the leg.

Osteomata form circumscribed hard nodular masses of bony consistency, and are usually painless. They may cause interference with function from their size, especially when they appear in close connection with a joint.

Treatment is complete and thorough removal.

Sarcomata are the most common tumors of bone; they are malignant, and when removed, tend to recur, either locally or by metastasis, in different parts of the body. The metastases usually are distributed by the circulation.

These tumors may arise from the marrow, but generally in the epiphysis of the bone and extend to the shaft only at a later stage of their development. As the tumor advances, it causes a softening and an absorption of the original cellular marrow until it approaches the periosteum.

In many cases the periosteum, as about any form of foreign body, then begins to proliferate and forms a shell of periosteal bone surrounding the tumor. In that way the shell of the bone oftentimes becomes very much enlarged before there is any extension of the process through the shell to the adjacent tissue. By destruction of the marrow and of the cortex, great softening of the bone may occur so that spontaneous fractures not infrequently are seen.

Other sarcomata arise from the periosteum, and usually originate from one side of the bone, although occasionally they entirely surround the bone. In the periosteal sarcomata, a new formation of bone is common and the bone is frequently arranged in a radical way, giving a most remarkable picture on the X-ray plate.

Myeloma is a very rare malignant tumor of bone. Such tumors always appear only in connection with bone, are usually multiple, and are of the same type as other lymphoid tumors.

The cells of such tumors resemble very closely the type of plasma cell. These cells are arranged in masses without an intercellular substance, and the tumors are closely allied to the malignant lymphomata. The cases are always associated with albuminuria.

Symptoms. The chief symptoms of malignant tumors are swelling and pain, both of which oftentimes are extreme. The swelling may be spherical or spindle shaped.

Extension to the joints may not occur for a great length of time. In many cases X-ray examination is the most reliable method of detecting the character of the bony change.

Treatment of all sarcomata is early and complete removal. This means in nearly all cases, amputation of the affected bone, and it is important that the amputation should be of the entire bone through the joint between the bone and the body, rather than amputation of the bone in continuity. The reason for this is, that even in sarcomata, which have not extended to the soft parts, very frequently there have occurred metastases of tumor-cells throughout the blood sinuses of the affected bone, often times at a distance of several inches from the site of the original primary tumor.

Carcinomata. Cancer of bone always is secondary to cancer in some epithelial organ. The infection may take place by direct extension through the blood or the lymphatics.

In cases of metastatic invasions of bone, spontaneous fractures oftentimes are the first symptom which calls attention to the fact that metastases have occurred.

Treatment. As in other malignant tumors, the indication is for absolute and radical removal whenever possible. Unfortunately, this very seldom can be done, because at the time the bone has become affected by extension to any great degree, radical operation is impossible. Many times, however, extensive operations must be undertaken for the removal of bone.

Cysts of Bone are rare lesions which practically always occur secondary to other lesions. They may occur as the result of the degeneration and softening of bone sarcomata. Some of the cases of bone cysts undoubtedly represent the entire destruction of sarcomatous processes. Occasionally echinococcus cysts of bone occur.

Treatment. Cysts of bone due to softening of the centre of sarcomatous tumors, like sarcomata themselves, are to be treated by complete removal, best usually by amputation. Cysts of bone not due to the presence of sarcomatous tissue, should be opened and drained in some cases. Cysts due to the presence of echinococcus, should be opened and drained, with the removal of every vestige of the echinococcus.

CHAPTER XI

DISEASES AND INJURIES OF THE ARTERIES AND GANGRENE

Gangrene is a term employed to denote the death of a part of the body, in mass.

Necrosis and *mortification* are terms used in a similar sense though necrosis is reserved in surgery to mean death of bone.

Gangrene may result from the gradual or sudden cessation of the arterial supply, or from a stoppage of the venous outflow. In general the etiology of gangrene comprises:

1. Traumatic causes.

2. Constitutional causes.

3. Thrombosis and embolism.

4. Cold.

5. The effect of certain chemicals.

Before entering into a consideration of these subjects, it is wise to first consider the varieties of gangrene.

There are two forms in which gangrene is observed: *dry* and *moist*.

Dry gangrene, or mummification, is a condition which occurs in consequence of a gradual diminution and final cessation of the blood supply, with the venous outflow intact. In this way, aided by evaporation and the venous return, there is a gradual drying of the parts. Diseases of the arteries and increasing pressure upon them from growing tumors, causes this variety.

Moist gangrene is due to the sudden arrest of the arterial supply, or a similar obstruction to the venous return.

This is the variety commonly met with from crushing or cutting accidents; from the effects of carbolic and other acids; from cold; and from thrombosis and embolism.

A *thrombus* is a blood clot occluding the lumen of a vessel. An *embolus* is a loosened part of a thrombus or any other foreign substance, free in the blood stream, such as a drop of fat, an air globule, or a detached particle of tissue from growths in the heart or vessels. Any one of these may find lodgment in a terminal vessel, and plug it.

Moist gangrene therefore differs from dry gangrene in that the arrest of circulation takes place more or less suddenly when the tissues are suffused with blood.

The dry form of gangrene does not occur regularly in the diseases in which it might be expected, and though a true wet gangrene is not found, neither is the typical mummification.

Moist gangrene may occur in diabetes, in senility and in Reynaud's disease, and probably assumes this form on account of the sudden onset of inflammation in the part from some slight abrasion, or from weak heart action.

CAUSES OF GANGRENE

Traumatic. The sudden cessation of the blood supply to a part in consequence of a cutting or crushing accident, will obviously produce the moist form of gangrene. It is not essential that the part be entirely severed, or even nearly so, for if only the main artery is severed, gangrene will ensue.

The crushing or pressure upon a large vein will act similarly, owing to there being no outflow possible, back pressure will cause the total arrest of circulation in the part.

Constitutional Diseases. Certain diseases affect the lumen or calibre of the blood vessels, gradually diminishing and finally arresting the stream of blood carried through them.

In these diseases it would be logical to invariably expect dry gangrene. This does not regularly occur, for the reason just given, and the mere presence of a moist or dry condition therefore cannot be regarded as diagnostic.

In diabetes, either form may obtain, and a diagnosis can be assured by the discovery of sugar in the urine.

The thickened condition of the arteries leading to senile gangrene must be thought of and proven in aged subjects. Dry gangrene is the rule in arteriosclerosis.

Reynaud's disease, or synthetic gangrene, is due to a vasomotor spasmodic condition of the terminal vessels and is of central nerve origin. The tips of the toes and fingers, of both sides, are the most common sites, though the lobes of the ears, cheeks and tip of the nose may be affected.

A coldness of the parts, with mottling of blue and white, and a subsequent diffuse blueness, becoming darker and finally black, are characteristic signs of this disease, and the dry form of gangrene is usual.

Obliterating Endarteritis, is a condition in which the walls of an artery become inflamed and thickened, thus obliterating its lumen.

Thromboangiitis Obliterans is similar to the above and differs only in that a thrombotic growth occurs in an artery obliterating its lumen.

Thrombosis and Embolism. Thrombosis and embolism cause a sudden or gradual stoppage of the blood stream in a vessel, and in consequence, either moist or dry gangrene occurs, depending on the time required for the obstruction to become complete.

The stoppage of the outflow because of thrombosis in a large vein, will cause moist gangrene; the part being unable to drain, will, by back pressure, arrest circulation.

Cold. Frost bite causes gangrene of varying degrees. A small circumscribed patch of tissue may succumb, or an entire finger or extremity may be affected. The variety is invariably moist. The diagnosis is easily made from the history of exposure.

Chemicals. Carbolic acid, even in weak solution, often causes gangrene of a finger or toe, because of its frequent use as a wet dressing, and therefore should never be employed in this manner. Gangrene of a single part, (especially in a young subject), incident to a slight injury or infection, should always excite suspicion that phenol has been employed. Moist gangrene is the rule. The part presents a hard, shriveled, black appearance which is characteristic.

Weak solutions of other chemicals such as lysol, acetic acid, and potassium or sodium hydroxide, employed as a wet dressing, are also capable of producing gangrene.

Symptoms. (*Dry Gangrene*). Typical dry gangrene usually develops in the toes and the feet, and the principal symptoms which point to its advent are, coldness, numbness, pain and tingling in the feet and muscles of the legs. Persons about to be affected with dry gangrene often complain for months, before any local signs of gangrene are present, of severe burning pain in the feet at night when warm in bed.

A trivial injury, such as a bruise, the friction of the shoe, or the cutting of a corn, may act as the exciting cause of the affection. The part becomes congested and gradually assumes a dark purple color, finally becoming black and dry; it is insensitive, but the surrounding parts are congested and may be the seat of intense pain. The dead part becomes black, shriveled, and dry, and emits little odor.

Dry gangrene usually spreads very slowly; one or two toes may first be involved and the disease may gradually spread to the rest of the foot and the leg. There may be little fever at first, but if a large extent of tissue is involved, a certain amount of fever develops. During the progress of the disease, pain is usually present to a greater or lesser degree, sometimes being intense; this is accounted for by the fact that the nerves are usually the last structures to die.

During the course of the disease, the patient loses much sleep from continued pain, and becomes worn out and may die of exhaustion.

In dry gangrene there is usually no well marked attempt at the formation of lines of demarcation and separation, but in some cases, if the amount of tissue involved is small, say one or two toes, or a part of the foot, for instance, and if the patient's strength can be sustained, the line of separation forms, and the dead tissue may be cast off, leaving the bones exposed in the wound.

Moist Gangrene. When a part which has had its vitality seriously interfered with becomes gangrenous, pain, which may have been present, suddenly ceases, the part becomes insensitive, and the skin is cold, pale, and mottled purple, green, and red, and finally dark colored; blebs containing brownish serum form upon the surface; the wound, if one is present, assumes a grayish color, and an offensive discharge escapes from it; the dead tissue rapidly undergoes putrefactive changes. Coincidentally with these changes in the dead tissues, the living tissue in contact with it becomes red and swollen, and the separation of the dead tissue from the living is affected by an ulcerative inflammation, granulations from the living tissue lifting off the slough.

The patient, at the same time, if the gangrenous process involves any considerable extent of surface, exhibits the unconstitutional signs of inflammation (fever, rapid pulse, etc.) and, in some cases, if the septic infection is intense, may die from septicemia.

In both dry and moist gangrene, when the gangrenous process is arrested, the dead tissue is separated from the living by a process of inflammation; the living tissue, at its point of contact with the dead tissue, and for some distance from it, becomes red and swollen, and exhibits all the signs of acute inflammation. The line of contact between the dead and the living tissue is known as *the line of demarcation*, and the line of granulations which separates the dead tissue from the living, is known as *the line of separation*.

The separation of the dead tissue is affected by granulations, which spring up from the living tissue as a result of inflammation, and there is also a certain amount of pus secreted from the granulations. In moist gangrene, the lines of demarcation and separation are fairly well developed. In dry gangrene, on the other hand, these lines are usually imperfectly developed.

Early Diagnosis. From the foregoing it will be observed that gangrene is most common in those past middle life, and that its actual onset is only a stage in an insidious process. This may be either due to senility or to some constitutional disease. A slight abrasion alone is sufficient to set up a train of symptoms out of all proportion to the cause. In such a case, the operation of a small verruca or papilloma may be followed by a violent inflammatory reaction, with rapid extension into the entire foot or leg, resulting in gangrene.

Such cases have occurred, but could have been prevented if a proper survey of the field had been taken and would have saved the chiropodist much responsibility.

Before operating on subjects past middle life, it should be a routine practice to note the color and temperature of the foot, both in the dependent and horizontal positions. The *anterior tibial pulse* should also be felt for and its absence or intensity noted. A question to the patient as to diabetes or thickened arteries may also elicit valuable information. A very weak or absent anterior tibial pulse (the knack of feeling the pulse here must be acquired), or peculiar nodules about the nail grooves, are evidences of an encumbered arterial supply.

Extreme redness or blueness in a foot in the hanging position, and pallor when elevated, also indicate a similar condition, or one in which the valves in the veins are impaired.

It is in such conditions that the greatest care should be taken to avoid deep incisions except in the presence of positive indications.

Treatment. In general, amputation through healthy tissue is the rule in gangrene affecting any extremity through its entire thickness. The complete devitalization of even a digital phalanx requires that amputation be made beyond the next joint above.

In traumatic gangrene it is the rule to amputate immediately through healthy tissue when restitution of the injured parts is known to be impossible. In senile gangrene the appearance of the line of demarcation indicates the extent of the devitalized area and establishes the point of amputation beyond the next joint above.

Diabetic gangrene presents the peculiarity of a slow and steady advance, unless an unusually high amputation be performed. Thus, if the great toe is the site of the beginning of a true diabetic gangrene, amputation through the lower third of the thigh is indicated; otherwise the prognosis is very bad.

Inflammatory gangrene, or as it is more properly called *gangrenous cellulitis*, is a rapidly spreading infective process which destroys tissue as it advances. It is an acute suppurative process causing large sloughs. It is a form of cellulitis requiring drainage and disinfection.

Frost bite may involve tissues to any depth and to any surface extent. Lesions of circumscribed contour result in the sloughing away of the area involved and never require amputation.

In the event of a phalanx, toe, finger, foot, or hand being involved, the same rules as above laid down must apply. In this variety, however, it is important to allow sufficient time to elapse in order that the depth of the gangrenous process may be ascertained. Should the line of demarcation be apparent, after a few days the complete death of the tissues below is certain, and amputation becomes necessary. If, however, after a few days some slight bleeding or the appearance of a red point be apparent, the bone, and in all probability some tissue around it, is still viable. Haste in these cases should therefore be avoided.

CHAPTER XII

DISEASES OF VEINS

Varicose veins are unnatural, irregular, and permanently dilated veins which elongate and pursue a tortuous course. This condition is very common, and twenty per cent. of adults exhibit it in some degree in one region or another.

The causes of varicose veins are obstruction to venous return, and weakness of cardiac action, which lessens the propulsion of the blood stream.

Varicose veins may occur in any portion of the body, but are chiefly met with on the inner side of the lower extremity.

Varix in the leg is met with during and after pregnancy, and in persons who stand upon their feet for long periods.

It especially appears in the long saphenous vein, which, being subcutaneous, has no muscular aid in supporting the blood-column and in urging it on. The deep as well as the superficial veins may become varicose.

Varicose veins are in rare instances congenital; they are most often seen in the aged, but usually begin at the ages of twenty to forty.

A vein, under pressure, usually dilates more at one spot than at another, the distention being greatest back of a valve or near the mouth of a tributary. The valves become incompetent and the dilatation becomes still greater. The vein wall may become fibrous, but usually it is thin, and ruptures. The veins not only dilate, but they also become longer, and hence do not remain straight but twist and turn into a characteristic form.

Varicose veins are apt to cause edema, and the watery elements in the tissues cause eczema of the skin. When eczema is once inaugurated, excoriation is to be expected. Infection of the excoriated area produces inflammation, suppuration, and an ulcer.

The skin over varicose veins in the legs is often discolored by pigmentation due to the red cells having escaped from the vessel and then being broken up.

The tissues around a varicose vein become atrophied from pressure, and often a very large vein will be in evidence whose thin walls are in close contact with the skin, and in this condition, rupture and hemorrhage are probable. Varicose veins are apt to inflame and thrombosis frequently occurs.

Treatment. The treatment of varix may be palliative or curative, but whichever is followed, endeavor first to remove the cause.

In palliative treatment, attend to the general health, keep up the force and activity of the circulation, and prevent constipation. Recommend the patient to exercise in the open air and to lie down, if possible, every afternoon. Locally, in varix of the leg, order a flannel bandage to support the vein and drive the blood into the deeper vessels which have muscular support. (For technic, see chapter on bandaging).

The curative or operative treatment of varicose veins consists of performing a resection of the internal saphenous vein of one or two inches, near the saphenous opening into the femoral. This is known as the *Trendelenburg* method. About 90 per cent of all cases can be cured by this method. The operation can be performed under local anesthesia and presents no difficulties.

Another procedure is known as *Schede's* method. This consists of making a circular incision around the leg just below the knee joint, and in tying all the superficial veins thus exposed.

Mayo's operation consists of the total extirpation of the internal saphenous vein from the saphenous opening to the internal malleolus. A small incision is made high up, and at a distance of from 8 to 10 inches, a second incision is made, and in this manner the entire vein is removed by making several incisions.

The patient should remain in bed about three weeks following an operation of this kind and afterwards an elastic stocking, or an ideal bandage, should be worn for a considerable time.

Phlebitis, or inflammation of a vein, may be plastic or purulent in nature. Plastic phlebitis, while occasionally due to gout, or to some other constitutional condition, usually arises from a wound or other injury, from the extension to the vein of a perivascular inflammation, or, in the portal region, from an embolus.

Varicose veins are particularly liable to phlebitis. When phlebitis begins, a thrombus forms because of the destruction of the endothelial coat, and this clot may be absorbed or organized.

Suppurative Phlebitis is a suppurative inflammation of the vein, arising by infection from suppurating perivascular tissues (*infective thrombophlebitis*). It is most frequently met with in cellulitis or phlegmonous erysipelas, but there are a great many other causes.

A thrombus forms, the vein wall suppurates, is softened and in part destroyed, and the clot becomes purulent. No bleeding occurs when the vein ruptures, as a barrier of clot keeps back the blood stream. The clot of suppurative phlebitis cannot be absorbed and cannot organize.

Septic phlebitis causes pyemia, and the infected clots of pyemia cause phlebitis. The symptoms of phlebitis are pain, which is at once felt in the limb along the track of the inflamed vein, and tenderness along the same area; the overlying skin is red, hot, and tender, and the lymphatic nodes in the groin swell; there is marked edema, but the inflamed venous cords can be readily felt. The constitutional disturbance is marked; rigors and high temperature, 103°F. to 105°F. (remittent type), are followed by profuse sweats. The general condition, facies and anxiety, dry and parched tongue, delirium and general distress, at once directs attention to the infectious nature of the trouble. The leucocyte count will show a marked increase in the number of polynuclears.

Treatment. The treatment of phlebitis may be classified into preventive and curative, the latter being subdivided into (*a*), general or symptomatic, and (*b*), local or surgical.

The preventive treatment is summed up in the word asepsis. The influence of asepsis in the management of wounds has completely revolutionized surgical practice, and the old fatal types of pyemia and septicema have now practically vanished.

Septic and pyogenic phlebitis still remain as consequences of accidental wound contaminations and as a penalty for the neglect of surgical cleanliness.

Prophylatic measures, by the use of internal remedies which diminish the coagulability of the blood, such as Wright's citric acid treatment, are recommended for the prevention of thrombosis. Antitoxins have not proven to be of benefit in this condition.

The curative treatment may be symptomatic, local, constitutional, or surgical. The constitutional treatment is directed to the general cause, if possible, as in the gouty, rheumatic, syphilitic, and chloritic cases; beyond this, there is no specific treatment. The antistreptococcal and staphylococcal sera are usually prescribed in the septic forms, but thus far, more as a forlorn hope than with the expectation of accomplishing any definite results. The symptomatic treatment, on the other hand, is always indicated to diminish pain, to support and strengthen the circulation, and to favor elimination. The main reliance is to be placed upon the local treatment, combined with good nursing, appropriate food, and moderate stimulation.

The local treatment is summed up in the following indications: (*a*), immobilization and absolute rest of the affected limb; (*b*), elevated position of the foot of the bed or of the limb to favor the drainage of the venous current toward the trunk. The limb should be covered with cotton batting and bandaged, over a gutter-splint of cardboard, extending from the foot to the thigh, to immobilize the knee. In the superficial inflammations, with much redness and heat, an even layer of any of the kaolin mixtures may be applied between thin layers of gauze, like an antiseptic poultice, over the entire extremity, and especially over the inflamed parts. A saturated watery solution of 25 per cent. ichthyol, painted over the entire surface will also prove decidedly beneficial in cases complicated with lymphangitis. Unguentum Crede, mercurial ointment, and the so-called resolvent lotions have been tried, but none of these can compare in their beneficial effect with kaolin poultices, with or without ichthyol, or the liberal application of broad compresses, thoroughly saturated with a weak lead and opium lotion, which latter acts not only as a local astringent, but as a marked sedative. Immobilization and rest should be maintained for a month or more.

Operative Treatment. The operative treatment of acute septic thrombophlebitis has in view three indications, and the procedures adopted must vary according to these: (1) ligation of the vein between the thrombotic focus and the uninfected vein on the cardiac side, in order to obstruct the further advance of the infection, and thus prevent the entrance of septic emboli into the circulation; (2) removal of the primary focus of infection by direct incision into the veins, evacuation of the septic thrombus and drainage; (3) extirpation of the infected veins with the contained clot and septic contents.

CHAPTER XIII

SPECIAL FORMS OF INFLAMMATION

Syphilis is a chronic, infectious, and sometimes hereditary, constitutional disease. Its first lesion is an infecting area or chancre, which is followed by lymphatic enlargements; eruptions upon the skin and mucous membranes; affections of the appendages of the skin, (hair and nails); chronic inflammation and infiltration of the cellulo-vascular tissue, bones and periosteum, and later, often by gummata. This disease is caused by a microorganism known as the *spirochaeta pallida* or *treponema pallidum* of Schaudinn and Hoffmann.

Transmission of Syphilis. This disease can be transmitted (*a*), by contact with the tissue-elements or virus acquired syphilis, and (*b*), by hereditary transmission, hereditary syphilis.

The poison cannot enter through an intact epidermis or epithelial layer; an abrasion or solution of continuity is requisite for infection.

Syphilis is usually, but not always, a venereal disease. It may be caught by infection of the genitals during coition; by infection of the tongue or lips in kissing; by the use of an infected towel on an abraded surface; by smoking poisoned pipes, and by drinking out of infected vessels.

The initial lesion of syphilis may be found on the finger, penis, eyelid, lip, tongue, cheek, palate, nipple, etc. Syphilis can be transmitted by vaccination with human lymph which contains the pus of a syphilitic eruption or the blood of a syphilitic person. Syphilis is divided into three stages (1) the primary stage—chancre and indolent bubo; (2) the secondary stage—disease of the upper layer of the skin and mucous membranes, and (3) the tertiary stage—affections of connective tissues, bones, fibrous and serous membranes, and parenchymatous organs.

Syphilitic Periods. (1) period of primary incubation—the time between exposure and the appearance of the chancre, from ten to ninety days, the average time being three weeks; (2) period of primary symptoms—chancre and bubo of adjacent lymph glands;

(3) period of secondary incubation—the time between the appearance of the chancre and the advent of secondary symptoms,—about six weeks as a rule; (4) period of secondary symptoms—lasting from one to three years; (5) intermediate period—there may be no symptoms or there may be light symptoms which are less symmetrical and more general than those of the secondary period; it lasts from two to four years, and ends in recovery or tertiary syphilis; and (6) period of tertiary symptoms—indefinite in duration; the fifth and sixth may never occur, the disease being cured.

Primary Syphilis. The primary stage comprises the chancre or infecting sore or bubo. A chancre or initial lesion is an infective granuloma resulting from the poison of syphilis. The chancre appears at the point of inoculation, and is the first lesion of the disease. During the three weeks or more requisite to develop a chancre the poison is continuously entering the system, and when the chancre develops, the system already contains a large amount of poison.

A chancre is not a local lesion from which syphilis springs, but is a local manifestation of an existing constitutional disease, hence excision is entirely useless. The hard chancre, or initial lesion, never appears before the tenth day after exposure, it may not appear for weeks, but it usually arises in about twenty-one days. The lesion commonly appears as a round, indurated, cartilaginous area with an elevated edge, which ulcerates, exposing a velvety surface looking like raw ham; it bleeds easily, rarely suppurates, does not spread, and the discharge is thin and watery.

The bubo of syphilis is multiple, consisting of a chain of glands, freely movable, indurated, painless, small and slow in growth, and the skin over the bubo is normal.

A positive diagnosis of syphilis can be made when an indurated sore is followed by multiple indolent glands or buboes in the groin and by the enlargement of distant glands.

Secondary Glands. The symptoms are noticed from four to six weeks after the stage of the induration of the chancre, and may continue to appear at any time, up to twelve months. The most constant are certain eruptions on the skin, faucial inflammation, and enlargement or induration of the lymphatic glands; others are febrile reaction, pains in the back or limbs, swelling of the joints, iritis and falling out of the hair.

Tertiary Syphilis. These symptoms appear from one to two years after contagion and may continue to break out from ten to fifteen years, or more. The characteristic lesions are certain late eruptions on the skin, periostitis and nodes on the bones, and growths in the subcutaneous tissue, muscle, and viscera, especially the liver and spleen. These growths, in the viscera and other parts, which are so characteristic of syphilis in its later stages, are known as gummata. They consist of a substance like granulation tissue, with a varying proportion of cells. In early stages they are grayish, gelatinous, and transparent, but the cells undergo fatty change and caseation takes place, so that the centre becomes yellow, and the circumference develops into fibrous tissue, which contracts like a scar tissue. Sometimes gummata break down completely, and suppuration, with destruction of the tissues in which they are situated, takes place; thus caries and necrosis not infrequently follow nodes on the bones.

Treatment. Mercury is the drug of great benefit in syphilis. This can be administered either internally, by inunction, or by injection. Of all the preparations to be given internally, protiodide of mercury, in one quarter grain doses, three times a day, is to be preferred.

Inunction represents the most efficient way of administering the mercurial treatment, when the stomach is intolerant of drugs, or when administered by the mouth in full doses, they do not favorably modify the symptoms. The patient is instructed to take a warm bath, and the mercury is then well rubbed in over the inner surface of the forearm and arm and alongside of the chest for fifteen minutes. Either the oleate of mercury, 10 per cent., or the ordinary mercury ointment is commonly employed; the former is more clean, but less efficient. The rubbings should be done by the patient, should be made over a large surface of the body, and should be performed thoroughly; one dram (4.0) of blue ointment is rubbed in daily. For the injections, a 10 per cent. salicylate of mercury in olive oil is to be preferred; 10 to 15 minums of this solution is to be injected into the buttocks, three times a week. The dose is gradually to be increased until 30 drops are employed. Recently salvarsan (606) in 0.6, or 10 grain doses is given either intravenously or intraspinally. Neosalvarsan (914) is to be similarly given. The latter has the advantage in that sterile water is used, and that, as a rule, there is no reaction from its injection. Iodide of potassium in large doses (60 to 90 grains) three times a day, is also to be given.

Tuberculosis. Tuberculosis is an infectious disease due to the deposition and multiplication of the tubercule bacillus in the tissues of the body. It is characterized either by the formation of tubercules, or by a wide spread infiltration, both of these conditions tending to caseation, sclerosis, or ulceration.

A tubercular lesion may undergo calcification.

A tubercule is an infective granuloma, appearing to the unaided vision as a semitransparent mass, gray in color, and the size of a mustard seed.

The microscope shows that a tubercule consists of a number of cell clusters, each cluster consisting of one or of several polynucleated giant cells, surrounded by a zone of epitheloid cells which are surrounded by an area of leucocytes. Giant cells, which also form by coalescence of the epithelioid cells, are not always present. The bacillus, when found, exists in the epithelioid cells, and sometimes in the giant cells; it may not be found, having once existed, but having been subsequently destroyed. It is often overlooked.

In an active tubercular lesion, even if the bacillus be not found, injection of the matter into a guinea-pig will produce lesions in which it can be demonstrated.

A tubercule may caseate, a process that is destructive and dangerous to the organism. Caseation forms cheesy masses, which may soften into tubercular pus, may calcify, and may become encapsulated by fibroid tissue. Tubercular disease of the bones and joints have already been described in a previous chapter.

Treatment. Destroy the bacilli present and radically remove infected areas which are accessible. Incomplete operations are apt to be followed by diffuse tuberculosis.

Bier's venous or obstructive hyperemia is especially to be recommended in tuberculosis of the ankle joint (for technic, see chapter on Therapeutics).

Plenty of fresh air, good nourishing food and tonics are indicated as a routine treatment.

Tetanus. Tetanus is an infectious disease, invariably preceded by some injury. The wound may have been severe or it may have been so slight as to have attracted no attention.

The disease is commonest after punctured wounds or lacerated ones of the hands or feet, and before it appears, a wound is apt to suppurate or slough, but in some instances the wound is found soundly healed.

Tetanus is due to infection by a bacillus (first described by Nicolaier, and first cultivated by Kitasato), the toxic properties of which, absorbed from the infected area, poison the nervous system precisely as would dosing with strychnine.

Symptoms. The onset is usually within nine days of an accident. At first, the neck feels stiff and there is difficulty in swallowing, and then the jaw also becomes stiff. The neck becomes like an iron bar, and the jaws are rigid as steel. If the injury is on the foot, that extremity usually is found to be rigid. Opisthotonos is present and spasms are very marked. Swallowing in many cases is impossible. The mind is entirely clear until near the end, one of the worst elements of the disease.

Treatment. Careful antisepsis will banish it. Every wound must be disinfected with the most scrupulous care. Every punctured wound is to be incised to its depth and thoroughly cleaned and drained. Large doses of the bromide of potassium, at least sixty grains, should be given every four to six hours. Tetanus antitoxin should be given (5000 units), and repeated in twenty-four hours if no improvement is seen. Recently a saturated solution of magnesium sulphate has been given intraspinally, with very good results. In all suspicious cases, a prophylatic injection of tetanus antitoxin is to be recommended (1000 units).

Erysipelas. Erysipelas is an acute, contagious disease, characterized by a peculiar form of inflammation of the skin. It is caused by the streptococcus of erysipelas, which grows and multiplies in the smaller lymph channels of the skin and its subcutaneous cellular layers, and in serous and mucous membranes.

The disease is a rapid spreading dermatitis, accompanied by a remittent fever, due to the absorption of toxins, having a tendency to recur. It is always due to a wound. The involved area may or may not suppurate.

Symptoms. The onset is sudden, with a high fever, and at the time of febrile onset, spots of redness appear on the skin. These spots run together, and a large extent of surface is found to be red and a little elevated. This combination of redness and swelling extends, and its area is sharply defined from the healthy skin. The color at once fades on pressure and returns immediately the pressure is removed. In the hyperemic area, vesicles or bullae form, containing first serum and later possibly sero-pus. Edema affects the subcutaneous tissues, producing great swelling in the regions where these tissues are lax.

Treatment. Isolate the patient; asepticize the wound; and give a purge. If a person is debilitated, stimulate freely.

Tincture of iron and quinine are usually administered. Nutritious food is important. For sleeplessness or delirium, use the bromides; for light temperature, cold sponging and antipyretics. Locally, strict antiseptic treatment of existing wounds or other lesions; cold compresses to relax the skin; rest; elevation of the limb; and incisions, only if pus forms.

Where the disease is spreading, good results are obtained by spraying the affected surface with a weak solution of corrosive sublimate in ether, or painting the borders of the affected area with contractile collodion. The affected part may also be painted with a 50 per cent. ichthyol and water solution. Alcohol, Burow's solution, and a great many other liquid applications are recommended.

Antistreptococci serum is also to be recommended; an initial dose of 20 c.c. followed by doses of 10 c.c., as often as necessary, being the usual procedure.

Cellulitis. In cellulitis, redness of the skin is not very pronounced and is late in appearing, following swelling, and not preceding it. It is essentially the same condition as a mild form of erysipelas. Its spread is heralded by red lines of lymphangitis, ascending from a wound (infected), swelling of glands, and fever.

In slight cases, the lymphatics may dispose of the poison, and suppuration fails to occur. In severe cases septicema arises. Cellulitis is usually a result of infection not only with streptococci, but also with other pyogenic cocci.

Treatment. Incise and curet the wound and apply one of the wet dressings.

Actinomycosis. This is an infectious disease characterized by chronic inflammation, and is due to the presence in the tissues of the actinomyces, or ray fungus. At the point of inoculation arises an infective granuloma, around which inflammation of connective tissues occurs; suppuration eventually taking place. Inoculation in the mouth is by way of an abrasion of mucous membrane or through a carious tooth. The fungi may pass into the bones and joints, causing inflammation of the parts. The bones in actinomycosis enlarge and become painful; the parts adjacent are infiltrated and soften; pus forms and reaches the surface through fistulae and the skin is often involved secondarily. In actinomycosis the adjacent lymphatic glands are not involved.

Treatment. Free incision, if possible, otherwise incision, cauterizing with pure carbolic acid, and packing with iodoform gauze. Internally, large doses of iodide of potassium should be given, as this drug alone has cured many cases.

Trench Foot. This results from exposure to wet and cold in the trenches, and soldiers who were compelled to have their feet immersed in water for any length of time and were then exposed to cold, are afflicted with this condition. The symptoms are similar to frost bite and the prevention of frigorism (Trench Foot) is as follows:

adequate feeding; perfect circulation; moderate exercise; good general health; and warm clothing, which all tend to give the body its maximum power of resistance to cold.

It is obvious that anything that tends to impair the circulation and the nutrition of the tissues is favorable to the occurrence of frigorism. Tightness of the clothing of the extremities, such as tight boots, leggins, etc., is particularly detrimental. Heavy clothing and other equipment, by increasing fatigue, also has a predisposing influence.

With regard to the protection against cold water, it is necessary that the external covering should be impervious to and not affected by water. India rubber stockings, waders, and boots have been used by men working in water, not only as a protection against wet, but also against cold. The best results have been obtained by the use of a waterproof covering that can be worn inside the boot, not because it is the only, or even the best possible method, but because it appears to be the simplest and most practical. A waterproof top boot, so devised as to leave a fairly wide air space between the boot and the greater part of the foot, ankle, and lower part of the leg, would be more efficient and probably more convenient, provided the material used was soft and light, and did not interfere with movements. To obtain this result a new type of boot would be required.

The treatment of trench foot is similar to that of frost bite.

Motorman's Foot. This is a condition caused by occupation, and the symptoms found are usually those of a flat foot combined with enlarged veins. The chief complaint is that of pain in the calf of the legs, which is increased upon standing for any length of time. The treatment is that for flat foot and enlarged veins.

Chauffeur's Foot. This is a condition also caused by occupation. On account of the position assumed in driving an automobile, the tendons and muscles of the leg are usually affected and a tendosynovitis very frequently occurs. The symptoms and treatment have already been described. Rest is without doubt the best therapeutic measure.

Bicycle Foot is another occupational disease. The chief symptoms are those of cramps in the calves of the leg, and pains of a severe neuritic character.

At times the onset is very sudden, and the cramps are so severe that it is impossible to extend the leg without causing great pain. Flat foot is usually associated with the above condition. The treatment is rest and the administration of the salicylates for the relief of pain.

Bicycling is ordinarily a beneficial exercise for the foot muscles. When bicycle foot results from this exercise it is usually evidence that the bicyclist had an abnormal condition of his foot muscles and foot joints before he took up the exercise in question.

CHAPTER XIV

VERRUCA (WART), CALLOSITY, HELOMA (CORN OR CLAVUS)

DISEASES OF THE NAILS – INGROWN NAIL

VERRUCA OR WART

Definition. A verruca is a circumscribed overgrowth of all the layers of the skin, varying in size from a pin's head to a small nut. These growths may be single or multiple, and may come and go without any special reason. *Verruca plantaris,* or plantar wart, is observed on the sole of the foot; it may be single or multiple. It is very painful; it may be the size of a pea and is often mistaken for a callosity, from which it may be distinguished by the pain on pressure, and the tendency to bleed when the horny layer is removed.

Verrucae are probably contagious, but the pathogenic agent has not been isolated. They sometimes disappear spontaneously, and they will recur if their removal is not complete.

Treatment. Certain chemical substances (see "*escharotics*") destroy tissue and can be employed with safety only after much experience. These drugs when allowed to spread on the normal skin often occasion painful and persistent lesions. They must therefore be applied directly and sparingly to the growth itself and not be left in contact too long.

The daily removal of a thin layer is possible in this way without causing pain or erosion.

The chemical agents that are employed for the removal of verruca are notably nitric acid, acetic acid, monochloracetic acid, trichloracetic acid, nitrate of silver, sodium hydroxide and salicylic acid. The treatment with these drugs is alike in all cases, with the exception of the last three named.

The procedure, when using liquid acids is as follows: render the growth and the surrounding parts aseptic; by means of a tapering glass rod or a wooden toothpick, apply a drop of the acid so that it will spread over the growth only, making certain that every part of the outer surface has been treated. If pain becomes excessive, apply a neutralizing agent. Dress the part with a shield that is holed-out, so that when the foot-covering is in place there will be no pressure over the tissues treated. This treatment should be repeated every other day until there is sloughing at the base of the growth. The pocket produced is drained, and balsam of Peru or some other stimulant should be applied and held in place by an appropriate dressing. Five or six treatments will ordinarily suffice to remove the growth.

Many practitioners find nitrate of silver a serviceable remedy in cases of verruca. The pure stick, moistened, is gently applied to the surface of the growth, which later becomes blackened. The patient returns two days later when the scab, that will have formed, is removed and the original treatment is repeated. Ordinarily from six to ten such applications will suffice. Those who favor the use of salicylic acid for the removal of verruca, usually apply a 60 per cent. ointment of this drug, over the growth only, protecting the surrounding parts with collodion or gelatine. A holed-out shield is applied over the growth and an appropriate bandage is made to hold it and the ointment in place. The patient is advised to return at the end of ten days and, as a rule, when the dressing is removed, it will be found that the growth is sufficiently loosened to admit of removal by means of forceps and scissors.

Sodium hydroxide is used in these cases in a saturated solution. It is best applied by means of a wood toothpick, wound about with cotton, and should be used sparingly, much after the manner in which liquid acid applications are made and as above described. A slight stinging sensation indicates that the drug has penetrated the tissues near the nerve-endings in the underlying papillae. Such symptoms render it necessary to neutralize the sodium hydroxide. According to Dr. Joseph Renk of New York City, ordinary vinegar contains just the degree of acidity necessary to neutralize the action of the sodium hydroxide, without adding a new irritating element.

Verrucae may also be removed by the high frequency spark, or by electrolysis. Both of these methods are superior to cutting operations, but are equally as painful unless a drop of anesthetic solution is injected into the base of the growth, before treatment is commenced.

CALLOSITY

Definition. A callosity is a circumscribed thickening of the *stratum cornium*. The condition is usually acquired, occurring on parts exposed to intermittent pressure with counterpressure from an underlying bony prominence, as on the toes, soles, and heel of the foot, from ill-fitting shoes.

Callosities are dirty-yellow to brown in color; their extent depending upon the cause; they are thickest in the centre and pass gradually into the healthy skin. Sensation is usually lost, or at least diminished, over these areas.

They may interfere with movement and may have painful fissures and become infected, giving rise to abscesses, lymphangitis, gangrene, or erysipelas. Hyperidrosis is often associated with this condition.

Treatment. The permanent cure of callosities depends exclusively upon the removal of their causation. The position of the foot in the shoe may be faulty because of excessively high or low heels, causing callous skin to appear upon the weight-bearing surface. Occupations requiring constant standing, and deformities, also enter as causative factors which must be considered.

The palliative cure rests for its efficacy on the removal of the horny tissue down to, but not into, the papillary layer.

HELOMA

(Corn or Clavus)

Definition. A heavy thickening of the cuticle, usually caused by pressure, and producing pain by its own pressure on the tissues beneath.

Though the term heloma is rarely used outside of text books, there are very few who have not had an unpleasant acquaintance with this cutaneous affection, under the name of "corns." Heloma is undoubtedly the most frequent of all skin diseases.

Cause. The exciting cause of helomata is intermittent pressure combined with friction; while among the predisposing causes it is only necessary to mention the slavish adherence to fashion which lends all of us to wear stiff leather shoes, the contour of which bears little or no relation to the natural shape of the anterior portion of the foot. The pressure of the ill-fitting boot upon the toes, or, more strictly speaking, the pressure of the toes against the unyielding leather, in walking, soon occasions hypertrophy of the horny layer at the point of irritation, and in time a dense, conical, pea-sized or larger mass is formed. The apex of the cone presses downward on the sensitive papillae and causes the painful sensation which suggests a visit to the chiropodist.

Helomata are named according to characteristics which mark them. When the growth is indurated it is called heloma durum; when soft, heloma molle; when of the millet seed variety, heloma miliare; when blood vessels are numerous, heloma vasculare. Each of these varieties requires a different method of treatment.

Helomata are most frequently found on the outer surface of the little toes, but may occur upon the sole of the foot and even upon the palm, or plantar surface of the foot. Between the toes they often form from pressure of the opposing digits, caused by narrow shoes, and in this location they are softer and usually present a whitish, macerated surface.

The Prophylatic Treatment consists in wearing a broad-toed, though not necessarily a square-toed shoe.

If shoes were made fan-shaped, like the imprint of a bare-foot in the sand, instead of having the greatest width across the ball of the foot, they might look strange at first, but they would be comfortable for all time. Those then who care more for comfort than for style, as most of us falsely profess to do, would have both cornless and comely feet.

The Palliative Treatment of helomata consists of first softening the dense, hard, horny tissue, when it will exfoliate spontaneously, or be readily scraped away. This projecting callous portion of the heloma may be removed by cutting or scraping till, as nearly as may be, the surface is level with the plane of the adjacent skin.

In the soft variety found between the toes, or in the vascular ones, located in the arch on the inner border of the foot, where the skin is thin, no thick covering will be encountered.

A line or groove will be observed marking the circumference of any variety of heloma, and it is in this line that the operative attack must be made.

Helomata of the miliary variety, usually appear on the sole of the foot and are, as a rule, as numerous as they are small. The preferable treatment is to use a sharp, pointed knife in removing each one of the "seeds" separately.

A well pointed, narrow blade introduced here will find a plane of cleavage between the growth and the surrounding tissue, through which it is possible to dissect quite deeply without encountering blood. When the dissection reaches the papillary layer in the skin, as evidenced by the red color, further operative steps should cease.

In the treatment of soft and vascular growths it may frequently be preferable to employ disintegrating solutions from the beginning.

Repetition of the treatment, as described in verruca, every second or third day, will result in the gradual disintegration of the growth to its extreme depth, and prove more satisfactory than the radical operation.

Healing is rapid and with the use of properly shaped, and roomy foot-gear, recurrence should not take place.

It is evident from the nature of helomata, that any "cure," rubbed or painted upon the affected surface, can only cause the softening of a certain thickness of skin, and that no hope for cure is justified unless the careful and complete removal of the growth is accomplished and followed by the use of roomy foot-gear.

Radical Cure. The total excision of corns, while disabling the patient more or less for a few days, is in many instances justifiable. There is little probability of recurrence if proper foot-gear is worn, and the results are especially good if the skin graft operation as devised by Dr. Robert T. Morris is employed, which is described in the next paragraph.

After the excision of the growth, a small piece of skin is removed from the leg and sewn to the denuded area. This prevents a tough cicatrix forming and assures a normal skin covering to the area previously occupied by the corn.

The Text Book of Practical Chiropody, now in course of preparation, will contain lengthy and explicit articles on the subjects of verruca and heloma. The purpose here has been largely to present the subject from a broad surgical viewpoint. The strictly chiropodial features will be thoroughly outline in the Text Book of Practical Chiropody after a manner never before attempted and will include all details of the chisel methods, the dissecting methods and the shaving operations.

DISEASES OF THE NAILS

INGROWN NAIL

Although chronic inflammatory affections of the neighboring skin often produce changes in the form, color and thickness of the nails, these so rarely call for surgical interference that only those conditions leading up to the development of ingrown nail will receive consideration in the following.

Ingrown nail may be due to either a lateral hypertrophy of the nail itself cutting into the soft parts, or to the primary hypertrophy of the soft parts themselves, thus producing the same picture. An accurate determination of which condition represents the original etiologic element is important in deciding upon a course of treatment directed to the radical cure of ingrown nail.

The term "radical cure" does not necessarily indicate the performance of the so-called radical operation, but may result from proper treatment of a down-curved nail edge, or of a diseased nail fold, together with such prophylaxis in foot-gear as is indicated. With sufficient room in the shoe and the removal of offending granulations or cutting nail edge, a radical cure can frequently be effected.

Any inflammatory condition, either of the nail or its matrix, or the tissues contiguous to the nail, may result in the train of symptoms which are indicative of ingrown nail. When, however, any of these conditions has existed sufficiently long to cause ingrown nail to be present, it ceases to be of the first importance; it then becomes necessary to treat the buried nail edge, or the overgrown soft tissues themselves.

The Choice of Method between radical and palliative operations will depend entirely upon the degree of infection present, and the facility with which it can be reached. Thus, in the event of the entire toe being red and swollen and much purulent discharge being present, there will in all probability also exist much inflammatory tissue and a deep burying of the nail edge.

With a tolerant patient it might be possible to scrape away with a sharp spoon the granulation tissue, and remove the offending nail edge; the gradual improvement sought in ordinary cases cannot be thought of in these cases. It is urgent to relieve the pain and throbbing and to circumvent the dangers of a spreading infection. The sensations of a cutting nail edge have been lost in the more severe development. Should the patient be tolerant of pain, exposure, disinfection and drainage of the infected area is possible, but in most instances the contrary will obtain, and the radical operation with local anesthesia will be indicated.

The possibility of doing an efficient operation will ordinarily determine the method to be employed.

On the other hand there are a large number of cases in which palliative treatment is not only effective but emphatically the method of choice. One might see a degree of burying of nail edge quite as extensive as in the foregoing, with however, only a slight degree of infection. The nail fold may be much hypertrophied and granulation tissue may be abundant. The tenderness and inflammatory condition, however, is not so great as to interfere with the ordinary procedure. There is no danger of a rapidly ascending infection, the nail groove showing no inordinate amount of discharge. It is in these cases that a permanent cure frequently results from the mere removal of the irritating nail edge followed by the disinfection of the nail groove.

It is held by many that all cases of ingrown nail, except those due to a true hypertrophy of the nail, would remain permanently cured were it not for short or badly shaped shoes.

The Palliative Treatment of Ingrown Nail must necessarily depend upon its original cause. Should it be due to the wearing of improper foot-gear, nothing primarily pathologic in the tissues themselves being present, treatment will be effective only when correct shoes are worn thereafter.

Eczematous skin surrounding a nail or infection of a nail groove or matrix, should be treated as such before sufficient hypertrophy takes place to bury the nail edge. The disinfection and drainage of the groove can usually be accomplished with iodin on a thin wire or wooden applicator inserted to the extreme depth of the groove, followed by the insertion of a narrow strip of gauze. Frequent changes of dressings and extreme cleanliness will cause the early subsidence of these infections. It, however, is to be deplored that in the early stages these cases so rarely obtain treatment.

Elevation of the nail edge is often practiced quite successfully, but in general, this method of treatment is not applicable to the acute stages of the disease on account of the concomitant pain. Either the nail is too thick to be elevated by the insertion of cotton under its free edge, or the soft tissues are too sensitive to admit of the pressure.

The real skill of the chiropodist is called into practice in the treatment of ingrown nail by palliative methods, and he may safely be judged by his results in this class of cases.

It requires discrimination whether to attack the exuberant granulation tissue or the cutting nail edge, and in many instances it will be found that both are necessary.

Much skill is required in removing that part of the nail which is buried without causing pain or bleeding; this is the first necessity for relieving pain and can only be accomplished by a technic acquired through practice, and often redounding more to the credit of the operator than the successful performance of a major operation. A sharp instrument, usually a chisel, is placed against the free edge of the nail so as to cut only through the nail itself and not into the nail bed, with the purpose in mind of removing a wedge-shaped piece of nail of just the size necessary to relieve irritation, and permit of proper drainage and dressing.

Exuberant granulations are best treated either with nitrate of silver applications (50 per cent.) or with tight packing, or both. Disinfection and wick drainage of the entire tract is of the utmost importance.

The Radical Treatment of Ingrown Toe Nail. The operations, as in the palliative treatment, naturally fall into two classes depending on (1) whether the nail originally was at fault, or (2) whether the soft tissues, by inflammatory processes, have hypertrophied and overgrown.

Operations depending on such diseases or malformations of the nail, causing it to grow down into the tissues, should be directed to the removal of the nail, or the offending part of it with its matrix. (See "*Hypertrophy*").

In conditions manifestly due to disease and hypertrophy of the soft tissues, palliative treatment frequently fails, and it becomes necessary to curet the granulating nail fold or to erode it with chemicals.

The best and easiest operation to effect a permanent cure, where this condition obtains, is known as Weber's operation. This operation consists of the excision of an elliptical section of tissue just alongside of the offending nail border, without interfering with the diseased tissues themselves, and suturing the cut edges together in the long direction of the wound. The incisions are made to extend a little further back than the nail and as far forward as possible. They are about a quarter of an inch apart at the centre and meet at these two points. The depth of the section of tissue removed, if sufficiently great, leaves a diamond shaped cavity. When the edges of the wound are brought together the overgrown edge is pulled away from the nail and the further cicatrization of the wound contracting the soft tissues, assures an excellent result.

HYPERTROPHY

Hypertrophy can result only from hyperplasia of the papillae of the matrix, the thickening of the nail occurring at the base, front, lateral edges, or over its whole extent, according to the parts diseased. The nail may be evenly thickened or variously curved or twisted, while its structure becomes brittle, opaque and discolored.

Removal of the most projecting portions of the nail will reveal the papillae elevated far above the normal level of the matrix.

The change is slow and progressive, and when pronounced is usually permanent. The causes are not well understood; pressure, however, seems to be an exciting cause, this being more causative in the nails of the toes, especially those of the great and the little toe.

The old, whose epithelial structures tend to overgrowth, are more liable to hypertrophy of the nails than the young.

When attacking the fingers, beyond the blunting of the tactile sensibility and the deformity, no special trouble arises, unless painful cracks form from the splitting of the brittle nails. When affecting the nails of the feet, however, it is difficult for the patient to wear shoes, the pressure leading to inflammation of the adjacent soft parts and eventually causing typical ingrown nail.

Back pressure upon the matrix from a short shoe upon a thick unresisting nail, is frequently the cause of onychia.

Palliative Treatment of Hypertrophy. When the deformity seriously interferes with the wearing of shoes, or shows a tendency to cut into the lateral fold, it becomes necessary to establish normal dimensions either with the knife or drill.

The total removal of the nail; including the matrix, is the only permanent cure. Excision of the cutting edge of the nail, as in radical operation of ingrown nail, eliminates only that element of discomfort.

The thinning of the nail, by scraping or with the drill, can also be accomplished with sodium sulphide. A sufficient quantity of the sulphide is added to starch paste to make it swell; this, when applied (use a wooden applicator) to the thickened nail, will cause the nail to disintegrate. By touching the surface with the applicator, one can determine the depth of nail destroyed before washing off the excess sulphide.

Radical Treatment of Hypertrophy. When the thick nail has cut into the lateral fold and actual ulceration has occurred, it becomes necessary to remove the down-curved edge.

Under local anesthesia, an incision is made through the nail, a little to the side of the inflamed area, and is carried well back through the matrix. A curved incision, outside of the infected fold, meets the first incision in front and back of the nail. All the tissue between is removed in one piece, including the offending portion of nail with its matrix and the nail fold with all granulation tissue.

This wound may be brought together by catgut sutures, or may be allowed to heal by granulation.

This operation suffices to prevent further trouble at the nail edge, but does not prevent the discomforts due to a long, distorted, horny nail. Total removal of the nail with its matrix is the only radical cure. (See "*Local Anesthesia*").

Inflammation of the Matrix (Onychia). As a result oftraumatism in unhealthy individuals, inflammation and suppuration sometimes occur at the root of a nail and in the contiguous portion of matrix ("run-around"), and often stubbornly continue unless the loosened, sharp edge of the buried nail be carefully trimmed away from time to time, and a little iodoform gauze be employed to press back the inflamed tissues.

From lateral hypertrophy of a toe-nail the sharp lateral edge becomes imbedded in the lateral fold, or from improper lateral compression of the toes, the same portion of soft tissues is forced up against the margin of the nail. In either case, inflammation, suppuration, and ulceration ensue, resulting in the formation of red, exuberant, excessively painful granulations, constituting the condition called *ingrowing toe-nail*, though more correctly it should be termed "up-growing pulp." Sometimes both edges, or even the whole matrix, become involved, producing pain on any movement of the member.

When inflammation and ulceration of the whole matrix occur, especially where a finger is involved, the condition is termed *onychia maligna*, which attacks only those in depressed health.

Treatment. The palliative treatment suggested for ingrown nail is indicated for all inflammations of the matrix, as far as the disinfection or removal of the portion of nail producing irritation is concerned, but in onychia maligna the whole nail usually requires removal under local anesthesia, with destruction of the matrix by caustics, or by curetment

Chapter XV

TUMORS AND CYSTS

TUMORS

Definition. A tumor is a circumscribed mass of tissue made up of cells of the same kind as the tissue from which it grows.

There are two distinct types of tissue in the body: epithelial and connective, and therefore two types of tumors: the *epithelial tissue tumors* and the *connective tissue tumors*.

Tumors may also be classified as *typical,* and *atypical.* A typical tumor is one in which the cells are identical to those in the tissue from which it springs, and also has the same arrangement of cells. They may be of epithelial or connective tissue origin. The tissue is identical in all respects and the growth is benign. An atypical tumor is one of epithelial or connective tissue origin in which, though the cells are the same as those in the tissue from which it grows, their arrangement is quite different. They are malignant.

The most important classification of tumors is that into *benign* and *malignant.*

A *benign tumor* is one in which there is no tendency to rapid growth; the symptoms are purely local, and the general health is not affected, except indirectly.

On the other hand a *malignant tumor* is one which takes on a rapid growth with a tendency to infiltrate or adhere to surrounding tissues; recurs when removed, and is accompanied by great pain and a rapid loss of weight and strength. These are commonly known as cancerous.

Malignant growths are of two types, carcinomatous and sarcomatous, dependent upon the tissue from which they emanate.

The *carcinomata* spring from the epithelial type of tissue while the *sarcomata* emanate from the connective tissue type.

Origin. Tumors originate from many causes. Some are congenital and others grow in later life from an inherited tendency.

Any continued irritation which acts mechanically or chemically so as to maintain a constant, though slight, degree of undue vascularity of a part, such as the hot, rough stem of a clay pipe or a jagged tooth, favors the development of a malignant growth. Certain benign growths, such as warts or moles, are especially prone to malignant change. Age and sex also predispose to tumor formation.

Thus carcinoma is a rarity under thirty years of age; the mammary gland of the female is more liable to carcinoma than the male; while on the other hand the esophagus, lip and tongue of the male are more liable to attack.

The possibility of certain malignant growths being of germ origin is thought to be evident (though not yet proven) from many facts. The fact that where there are malignant growths present, lymphatic glands, quite distant from the original growth, become secondarily infected, through the lymphatic vessels, seems to carry out this view.

Particles of a carcinoma (metastasis) floating in the blood stream, finding lodgment elsewhere also establish new growths (metastatic).

Tumors are named according to the tissues from which they arise, thus:

Connective Tissue Tumors

Fibrous tissue— —Fibroma

Fatty tissue— —Lipoma

Mucous tissue— —Myxoma

Muscular tissue— —Myoma

Cartilage— —Chondroma

Bone— —Osteoma

Blood vessels— —Angioma

Lymphatics— —Lymphangioma

Lymphatic glands— —Lymphoma

Epithelial Tissue Tumors

Warty— —Papilloma

Glandular— —Adenoma

Skin— —Epithelioma

CYSTS

Definition. Cysts are hollow tumors filled with fluid or semi-solid contents. They are classified according to their mode of development:

1. Cysts formed in already existing spaces such as sebaceous cysts in the sebaceous glands of the skin; mucous cysts in mucous glands, and distension cysts in ducts of large glands like the salivary, lacteal, hepatic, etc.

2. Cysts of new formation into the tissue spaces from the effusion of blood or plasma.

3. Congenital cysts known as dermoids.

4. Cysts of parasitic origin.

The only cyst with which the chiropodist ordinarily comes in contact is of the sebaceous variety.

Sebaceous Cyst. A sebaceous cyst is a tumor resulting from retained sebum (secretion of the sebaceous glands).

They sometimes, though rarely, are found on the soles of the feet. They range in size from a millet seed to the size of an egg or larger; they may be globular or flattened. They may be single or multiple; the skin over them is normal in color and smooth, or white if distended, red if inflamed. They grow very slowly and ordinarily persist indefinitely, but calcareous changes are common. Not infrequently they break down and ulcerate. The wall is made up of connective tissue lined with epithelium and the secretion if chemically altered, becomes fluid, semi-fluid, cheesy or purulent.

Treatment. Spontaneous cure often occurs when a cyst becomes inflamed and suppurates. The pus is evacuated either spontaneously or by incision, following which the walls of the sac adhere and its cavity is obliterated.

Treatment directed toward the obliteration of the sac is the only procedure which gives promise of permanent cure; mere puncture and evacuation will effect only temporary relief, the sac soon filling again.

Incision followed by dissection and removal of the sac, either intact or punctured, is radical and efficient.

Puncture and evacuation, followed by swabbing out with pure phenol or strong iodin, may set up an inflammatory reaction within the sac, which acts similarly to the suppurative process, causing adhesion of the walls, thus preventing a recurrence.

CHAPTER XVI

FRACTURES, DISLOCATIONS AND SPRAINS

FRACTURES

A fracture may be defined as a broken bone. Fractures are classified as follows:

1. As to their degree.

2. As to the direction of the line of fracture.

3. As to their location.

4. As to the etiology.

5. As to their relation to the overlying skin.

6. As to the number of fragments.

7. As to whether they are complicated or not.

Degree of Fracture. A fracture which only involves a portion of the thickness of the bones, so that its continuity has not been entirely lost or a fragment has not been completely detached, is called an *incomplete fracture*. A fracture which involves the entire thickness of the bone, so that it is divided into two or more distinct fragments, is called a *complete fracture*.

INCOMPLETE FRACTURES

Among the varieties of incomplete fracture are: greenstick; fissured; depressed.

Greenstick Fractures (really a bending rather than a break of the bone) are mostly seen under the age of fifteen, and the bones of the leg are rarely affected.

Fissured Fractures are those in which there is a split or crack in the bones; they are very rare in the bones of the lower extremity.

Depressed Fractures are fractures in which one or more segments of broken bone are depressed; they are most common in fractures of the skull.

COMPLETE FRACTURES

Complete Fractures are divided according to the line and the seat of the breech of bone continuity.

Directions of the Lines of Fractures

Transverse, when the line of fracture does not deviate more than ten to fifteen degrees from that of the transverse axis. This variety is rare in the shaft of the long bones. It is usually found at the lower end of the radius or of the femur, and in the short bones.

Longitudinal, when the break is parallel to the long diameter of the bone; very few cases of this variety are seen.

Oblique, when the direction of the line of fracture may form any angle with the transverse axis of the bone up to a right angle. When it approaches the latter, it belongs to the group of longitudinal fractures. In the oblique variety, the line of fracture may be single or multiple. This and the spiral form are most frequent in the shafts of the long bones.

Spiral, when the break line is spiral. This variety of fracture was formerly considered to be very rare. The more systematic use of the X-ray as part of the routine of diagnosis has shown that spiral fractures are quite frequent in the shafts of the tibia and fibula. They are usually the result of a rotating or twisting force.

Classification of Fractures

Comminuted, when there is extensive splintering of the bone adjoining the fracture or one of the fragments.

Impacted, when the fragments are driven into each other. This variety usually occurs in the neck of the femur.

Compression, or Crushing Fractures, when the broken bones are compressed or crushed; this variety usually occurs in the tarsal bones. The spongy portion and cortical layer are both crushed. In some cases there is a perfect pulpification of these bones. This condition occurs after falls from a height upon the sole of the foot.

Location of Fracture

In the Diaphysis of a Bone. Breaks in the diaphysis of a bone are spoken of as fractures of the *shaft,* and to be still more exact, it is stated whether of the upper, middle, or lower third.

At the Ends of Bones. Fractures occurring at the ends of bones receive the name of the part which the line of fracture transverses; for example, fractures of the *neck* of a bone, of a*tuberosity,* of a *process,* of a *condyle,* etc.

There are two forms of fracture that require special mention in connection with their location. These are *epiphyseal separations* and *articular fractures.*

Epiphyseal Separations. The union of the epiphysis to the diaphysis commences during puberty, hence these fractures are less common in childhood than after the ages of eleven or twelve. As a rule, they can only occur before the twentieth year. The periosteum is more resisting and tougher during the early years of life than later on.

Articular Fracture (*joint fractures*). Like epiphyseal separations, recognition and proper treatment of these fractures have assumed great importance.

Articular fractures may be divided into three classes:

1. *Intra-articular.* In these the line of fracture lies entirely within the joint. Such fractures are most frequently found in the elbow and knee joint.

2. *Para articular.* In these the line of fracture extends close to the joint but not into it. An example of this class is the *supracondyloid* fracture of the humerus.

3. *Articular fractures proper.* The majority of joint fractures belong to this class. The line of fracture either extends into the joint from without or it extends from the joint outward. As example, the ankle

joint; the majority of the typical supramalleolar, malleolar, and spiral fractures of the tibia and fibula.

Etiology. Fractures may be divided into two groups: the *traumatic* and the *pathologic* or *spontaneous*. In the traumatic, the fracture is the result of violence acting upon a bone which is either normal or shows slight changes due to the physiologic causes mentioned. A pathologic or spontaneous fracture is one which occurs in a bone, the strength of which has been diminished by some preceding abnormal or pathologic changes. In this variety the degree of force which produced the fracture would not be sufficient to cause a fracture in a healthy bone.

The causes of traumatic fractures may be either predisposing or exciting.

Predisposing Causes. The bones of the human body attain their greatest strength toward middle age. From infancy up to that time the bones are very elastic and yielding. Toward old age an interstitial atrophy occurs. It causes a thinning of the cortex of the shafts and of the trabeculae of the spongy portions of the long and short bones. It is an actual diminution of the bone substance and a corresponding increase of the fat. This is especially seen in the neck of the femur. When it occurs in old age, it acts as a predisposing cause, but when it occurs prematurely or reaches an extreme degree, it must be considered as pathologic.

Existing or Determining Causes of Fractures

Fractures by External Violence are divided both clinically and from a mechanic standpoint into two classes: *direct* and *indirect*. In fractures by direct violence the bone breaks immediately under the point where the force has been applied. In this class of fractures there is more damage to the soft tissues and this damage is generally more serious than in indirect fractures. Direct fractures are more likely to occur in exposed bones like the clavicle, os calcis, etc.

An example of fracture by direct violence is found in fractures of the tarsal bones after a fall upon the feet from a height.

Under the head of fractures by indirect violence belong (a) those which occur as the result of a rotary or twisting force (spiral fracture of the tibia or fibula, for example); (b) those which are produced by compression; (for example, a fall upon the feet may cause an impacted fracture of the upper end of the tibia); (c) those which are the result of a tearing force.

Fractures resulting from a tearing force occur when a joint is suddenly moved beyond its normal range of excursion. The firmly attached ligaments being a fixed point, the ends or some process of the bones composing the joint are torn off from the remainder of the bone. Examples of this are fractures of the internal or external malleoli, following forcible eversion or inversion of the foot.

Fractures are also caused by muscular action and by gunshot injuries.

Pathologic (spontaneous fractures):

1. Fractures resulting from bone fragility of local origin as for example, tumors, osteomyelitis, aneurisms.

2. Fractures resulting from bone fragility due to some general disease, as for example, tabes dorsalis, paresis, rachitis, osteomalacia, and exhausting chronic diseases.

Classification and Relation of Fractures to the Overlying Skin

Fractures are divided into *compound,* or *open* and *simple,* or *subcutaneous,* according to whether a communication does or does not exist between the seat of fracture and a wound of the skin.

A compound fracture is one in which the cutaneous wound communicates with the seat of the fracture.

A simple fracture is one in which a wound of the skin is absent, or, if present, no communication exists between it and the seat of the fracture.

The majority of compound fractures are the result of direct violence, and the injuries of the soft parts, are, as a rule, far more extensive and serious than in a simple fracture. A fracture which is simple at first, may become compound as a result of necrosis of the skin lying over it; or as a consequence of the original injury; or of pressure upon it by a displaced fragment; or by penetration of the skin, in efforts to use the limb.

Further Classification of Fractures

Fracture. In the ordinary use of the term "fracture" is understood to indicate a *complete* or *incomplete* separation of the bone into two or more fragments, the lines of which are continuous with each other.

Multiple Fracture. The term *multiple fracture* is applied to the simultaneous fracture of two or more non-adjacent bones, and also to those cases in which two or more fractures of the same bone exist, and the lines are not continuous with each other. Such multiple fractures are usually the result of direct violence.

Complicated Fracture. When a fracture is accompanied by injuries of the viscera, nerves, etc., the term *complicated fracture* is applied. Such a fracture may be simple or compound. The term complicated, as ordinarily employed, is limited to those fractures which are accompanied by local, rather than by general complications.

Symptoms of a Recent Fracture. In the examination of a patient who has sustained a recent fracture, procedure should be as follows: the history of the patient and of the accident should be taken; an examination should be made for objective signs, like deformity, abnormal mobility, crepitus, and ecchymosis; subjective symptoms, such as pain and loss of function of the limb should be ascertained; an X-ray picture should be taken and every possible precaution observed to exclude distortion or exaggeration.

Treatment of Fractures. *First Aid.* The treatment of fracture may be said to begin from the moment of its occurrence. Much can be done for the comfort of the patient and correct union of the fracture by intelligent treatment during the first hours.

The proper temporary fixation of the limb, the mode of transportation, and the removal of the clothing, all require special mention.

The use of first aid dressings, those which can be used until more permanent and suitable ones can be applied, varies, of course, with the individual bone affected. In fractures of the tibia, fibula and foot, as well as in those of the lower half of the femur, the use of the blanket splint will be found of great aid. Instead of a blanket, a long pillow or soft cushion can be employed in the same manner.

The "blanket splint" can be readily made by folding a blanket in such a manner that it extends from the middle of the injured thigh to below the foot. Two pieces of narrow, strong board, or better still, two broomsticks are rolled up in the blanket, one at either end. The rolled-up blanket is now turned in so that the board supports with their enveloping turns of blanket, lie upon the posterior surface. Thus, a trough is formed in which the limb is placed and firmly secured by loops of bandage, one below the foot, the second just above the ankle, the third below the knee, and the fourth near the upper end of the blanket.

In fractures of the leg, after the application of the emergency splint, the patient should be transported in a recumbent position, the support being as firm as possible, a wide board, shutter or a wooden rail being preferable. If such supports are not at hand, and the patient is to be moved without their use, the persons transporting the invalid should be distributed in the following manner: one supporting the head and shoulders, a second the pelvis, and the third the two limbs.

Reduction. The reduction of a fracture is the effort made by the surgeon to overcome any tendency to displacement, and thus to place the fragments in such close apposition that an accurate and firm union is possible. The best time in general for the reduction of a fracture is as soon as possible after the accident, if the patient's general condition will permit. If there is marked displacement of fragments, so that there is danger of necrosis of the overlying skin or of damage to the adjacent vessels or nerves, an early reduction is imperative.

In all cases in which reduction is very painful or difficult, whether performed shortly after the accident or at a later period, it is best to administer an anesthetic to overcome muscular contraction and to decrease the amount of pain. After reduction of a fracture, retentive apparatus is indicated in order to maintain apposition. In the use of dressings there will be two kinds, those which are temporary and those which are permanent. The former are employed where the swelling of the limb is such that some dressing can be employed which will not cause pressure.

Certain general principles should be followed in the use of splints; for instance, a splint, after being applied, should not interfere with the circulation, allowance always being made for the swelling of the limb, which almost invariably occurs during the first week. The splint, if flat, should be wide enough to obviate the possibility of pressure against the point of fracture; also, it should project a little beyond the limb.

In general, it is best to immobilize the adjacent joints, above and below the seat of fracture, but no dressing should be permitted to remain so long as to produce stiffness of the joints and muscular atrophy.

The skin, even in simple fractures, must be cleansed with green soap, water and alcohol. If blebs or an area of threatening necrosis of the skin exist, they should be freely dusted with powdered boric acid and a few layers of aseptic gauze applied.

The form of retentive apparatus to be employed will vary, of course, with the individual bone requiring treatment.

The most important articles of a fracture equipment are as follows:

1. Plaster of Paris bandages for making molded splints and circular casts.

2. A stock of basswood, three-sixteenths of an inch thick, for making wooden splints.

3. An assortment of metal splints or materials for making them.

4. Muslin for bandages and slings.

5. Five yard rolls of ordinary and zinc oxide adhesive plaster, three inches wide.

6. Cotton batting and sheet wadding for padding splints.

7. Strips of tin or thin cypress for strengthening plaster casts.

The selection of a dressing for the immobilization of a fracture depends upon, *first,* the particular bone involved and whether apposition can be maintained with or without extension; *second,* whether great swelling be present or not; *third,* whether the fracture be simple or compound; and *last,* whether ambulatory treatment be preferable to that in the recumbent position. This latter applies, of course, only to fractures of the lower extremity.

Operative Treatment of Simple Fractures. Operative treatment of a recent simple fracture is indicated in general, when reduction cannot be completely made; when correct apposition cannot be maintained; when there is interposition of bone or soft parts; when the fracture is a spiral one with considerable displacement of the fragments; when fragments are rotated upon each other, and when there are multiple fractures.

The most favorable time to operate in recent simple fractures is at the end of the first or beginning of the second week. At this time the process of callus formation is most active. The blood clots and loose shreds of tissue have begun to be absorbed, so that the fragments are more easily accessible.

Methods of Fixation of the Fragments. In the majority of cases the reposition of the fragments alone is not sufficient to maintain accurate apposition. It is usually necessary to employ some means of mechanical fixation. In all the methods employed, the preparation of the parts is the same as for any aseptic operation. The opportunity for serious complications resulting from septic infection, is greater than in any other class of operations. It is for this reason that extraordinary caution must be exercised. The incision should be large enough to expose the seat of the fracture thoroughly.

The materials used to secure fixation are: absorbable sutures, such as chromicized catgut or kangaroo tendon; metal suture of silver or bronze aluminum wire; screws, nails, plates, clamps, etc.

Injuries in the Vicinity of the Ankle Joint. In the examination of a patient who shows evidence of injury in the vicinity of the ankle joint, such as swelling, deformity, loss of function, etc., the following conditions must be thought of, in the order given:

1. Fractures of the lower ends of the tibia and fibula (Pott's Fracture).

2. Dislocation at or near the ankle.

3. Fractures of the tarsal bones.

4. Rupture of the tendon Achillis.

5. Sprains of the ankle.

Fractures of the Lower Ends of the Tibia and Fibula. Commonly given the name of *Pott's Fracture*. They may be the result either of forcible abduction or eversion of the foot, or of inversion or adduction. If the sole or main movement is eversion, the *internal* malleolus is broken, and if the force continues to act, it also causes the *external* malleolus to be broken. In the second variety, fracture by inversion, the first effect of the force is to break the fibula at the external malleolus. If the movement continues, the internal malleolus or a greater portion of the tibia is broken off.

Diagnosis. The diagnosis is usually easy to make. The ankle joint is greatly swollen, the depression, normally present in front of and behind the malleoli, being obliterated. The foot is displaced outward, and the internal malleolus is prominent. This deformity will often persist and become a cause of disability after healing of the fracture.

There is also backward displacement of the foot. These displacements may be so marked as, at first glance, to resemble a true dislocation of the ankle.

Abnormal lateral and anteroposterior mobility may be ascertained by grasping the sole of the foot with one hand and moving it inward and outward, or backward and forward, while the other hand steadies the leg. There is great tenderness between the tibia and fibula at the front of the ankle, and over the points of fracture in the malleoli.

If the fibula alone be broken, abnormal mobility and crepitus may be elicited by pressing its tip inward with the index finger of the one hand while a finger of the other hand is placed at the seat of fracture.

In some cases of Pott's fracture the foot will move inward instead of outward. The degree of outward displacement can be measured by the difference in the distance from the front of the ankle to the cleft between the first and second toes, as measured on the sound and injured foot. There is not always complete loss of function. In fractures of the external malleolus alone, the patient may walk quite well.

Treatment of Fractures of the Leg. The treatment of a simple fracture of one or of both bones of the leg depends *first,* upon whether or not swelling is present, and *second,* upon the amount of displacement of fragments and our ability to keep them in apposition after reduction. If the case is seen within a few hours after the injury and but little, if any, swelling be present, the following is a perfectly safe and justifiable method of treatment:

The limb is wrapped with strips of sheet-wadding from the toes to the middle of the thigh, and a circular plaster of Paris cast is applied extending over the same area. Before the cast is dry, it is cut open along the median line, in front, to allow for any swelling. The cast is best applied while the patient is under the influence of an anesthetic, so as to permit reduction of the fragments by traction upon the foot. In from ten days to two weeks the cast should be removed and a fresh one applied. The second cast does not require to be cut open, and can be left on the limb until the end of the fourth week. It is then removed and if union be complete, no further cast need be worn. Massage of the limb and passive and active motion are now begun.

Fractures of the Tarsal Bones. Fractures of these bones have been found far more frequently than was thought before the use of the X-ray. Many cases of tarsal fracture have been treated for sprains of the ankle. It is only when the recovery is slow or the injury is followed by a traumatic flat foot that the surgeon begins to suspect that a more serious condition was present at the time of the original injury.

The astragalus and os calcis are the tarsal bones that are usually affected. Fractures of the os calcis, in the majority of cases, are due to compression. The patient falls from a height to the ground, on a hard substance. The os calcis is crushed between the astragalus and the ground.

There are three general types of fracture of the os calcis:

1. That in which the fracture has been confined largely to that portion lying behind a vertical plane through the middle of the body of the astragalus. There are three varieties of this heel fragment type: (*a*) cases with one large heel fragment; (*b*) cases of small heel fragments (in this variety, also called avulsion fracture, the sudden contraction of the calf muscles pulls the fragment off; at times the tendo Achillis itself is torn off from the attachment to the os calcis at the same time); (*c*) cases showing only fissures in the bone.

2. Comminution of the anterior half of the os calcis.

3. All the cases of extensive comminution of the bones; the bone is literally shattered.

Fractures of the Astragalus. These can be divided into: (*a*) those of the neck; (*b*) those of the body. The former are the most common fractures of the astragalus. They may follow sudden dorsal flexion, or forced supination, or pronation of the foot. They may be due to a fall from a height or from direct violence. Fractures of the body of the astragalus are usually the result of a crushing force which ordinarily have a like effect on the body of the os calcis, and are often associated with fractures of the latter bone. The variety of fractures is considerable, varying from two large fragments, to complete comminution of the bone.

A fact of considerable importance in the interpretation of skiagraphs of fractures of the astragalus, is a knowledge of the presence in many normal individuals of a little bone known as the *os trigonum*. It may occur detached from the astragalus or may be attached to it as a process, on its posterior aspect, and on account of the swelling and pain around the ankle, a diagnosis can seldom be made without the routine use of the X-ray in every injury in this region.

The swelling, with obliteration of the depressions normally present around the ankle, does not differ from that characteristic of a sprain of the ankle or of a Pott's fracture. If there is extensive comminution of the os calcis or astragalus, the malleoli may be a little lower than normal.

The X-ray must always remain our most reliable means of diagnosis at the time of the injury. At a later period the chief symptoms are a painful flat foot, ankylosis of the ankle joint, pain and difficulty in pronating and supinating the foot.

The prognosis of fractures of the tarsal bones is not favorable, even though the lesion has been recognized at the time of injury. Even in the most favorable cases there is some limitation of lateral motion. The outlook is better in those cases of fracture of the os calcis in which there is a large heel fragment, than if the fracture is comminuted. The most frequent sequel is stiffness of the ankle-joint and traumatic pes valgus. Infection is frequent in compound fractures.

Treatment. This does not differ from that of a Pott's fracture until the greater part of the swelling has disappeared. The skin of the foot and lower portion of the leg should be thoroughly cleansed and covered with gauze. This is necessary on account of the possibility of necrosis of the skin of the heel, and the danger of infection of the bruised soft tissues around the heel.

The foot should be placed in a well-padded box or in a posterior splint of the Volkman type. Ice bags should be applied over the sides of the heel.

After from eight to ten days, a circular plaster cast can be applied, extending from the toes to the knee. An anesthetic should be given during the application of the cast, the foot being held flexed at right angles and sheet wadding freely used around the ankle. The cast should be worn for seven weeks. At the end of this time the patient is gradually permitted to step upon the injured foot. Passive and active motion are also now employed.

Fractures of the neck of the astragalus, with rotation of the posterior fragment, are usually followed by great limitation of the movements of the ankle joint. This condition might be greatly improved by an open operation.

Fractures of the Metatarsal Bones. These are usually due to direct violence, as occurs when a heavy weight falls upon the dorsum of the foot. Another example of direct violence is a fracture following a crushing injury, as in being run over.

In indirect violence, such as follows dancing, jumping, or sudden twists of the foot, the fifth metatarsal bone is the one most often involved. There is but little tendency to displacement except when several bones are broken at the same time, and then it is toward the dorsum of the foot.

The diagnosis in fractures produced by direct violence is made from the following: presence of severe localized pain; swelling; and, not infrequently, crepitus and abnormal mobility. In those fractures due to indirect violence (second, third and fifth metatarsals), there is pain when the patient endeavors to put pressure upon the toes or tries to invert the foot. The usual signs of fracture are absent. A skiagraph should be made in every case.

Fracture of the metatarsal bones is liable to be followed by traumatic flat foot, on account of the sinking of the arch, or painful large calluses forming on the sole of the foot may interfere with walking.

Treatment. The treatment in such fractures is by immobilization in a posterior metal or plaster splint, for four weeks. If there is continual pain upon walking after the injury, a steel insole will often give relief. The treatment of compound fractures of the metatarsal bones does not differ from that of other bones.

Dislocations. A dislocation is a displacement from each other of the articular ends of the bones which enter into the formation of a joint. A diagnosis can usually be made from certain objective and subjective symptoms, taken in conjunction with an accurate history of the manner in which the accident occurred.

Examination should be made in a systematic manner in every case, us follows:

(1) *Inspection.* The limb should be first inspected to note the position, the alterations of contour, or of the axis of the limb, or the projection or absence of certain bony prominences. The position is often so characteristic that a diagnosis can be made by inspection alone.

(2) *Palpation.* By this one can learn the relation of the displaced articular ends to each other, unless the swell ing is too great, or the patient is very stout. This method also enables one to ascertain the absence of normal prominences or the presence of abnormal ones. The end of the displaced bone may be felt in an abnormal position.

(3) *Measurement.* The limb may only appear to be or is actually shortened. In the latter event the normal measurements between bony prominences will be altered.

(4) *A skiagraph* should be made in all doubtful cases to confirm the diagnosis of dislocation, and also to ascertain whether there is an accompanying fracture.

When the patient is stout, or when considerable swelling exists the use of the X-ray is of especial value.

The attitude of the limb is often so characteristic that simple inspection will enable one to make a diagnosis by this means alone. In stout persons, a change in the axis of the limb or a change in position is apt to be overlooked. The relation of the articular surfaces can be determined by palpation, unless the swelling is too great. Measurement of the limb will usually show a shortening, depending upon the position in which the limb is held. The movements of a dislocated joint are usually limited. If any movement of the end of one of the bones is felt, it is always at an abnormal point. Pain is referred to the dislocated joint and the patient is unable to use the limb.

Treatment. As a rule, a dislocation should be reduced as soon as the diagnosis is made, and, if necessary, an anesthetic should be administered.

When reduction has been accomplished, the bone often goes back with a snap, the contour of the limb is restored, and the movements of the joint are free again.

If it is impossible to reduce a recent dislocation, the following obstacles must be considered: (*a*) interposed portions of the capsule; (*b*) interposed muscles or tendons or sesamoid bones; (*c*) torn off fragments of bone; (*d*) a fracture of the shaft close to its articular end, which would prevent its being used as a lever for reduction.

The after-treatment of a dislocation is usually quite simple. A bandage or splint should be applied, which will keep the joint immobilized for a period of two weeks, after which passive motion and massage can be begun for fifteen minutes twice daily, the splint or bandage then to be reapplied for another two weeks.

DISLOCATIONS AT THE ANKLE JOINT

Backward Dislocations occur more frequently than those in a forward direction.

The injury usually is the result of a fall backward while the foot is flexed. This causes an extreme plantar flexion of the foot. The astragalus, and with it the foot, is displaced backward. The lateral ligaments are usually extensively torn. In the majority of cases there is an accompanying fracture of either one or both malleoli or of the shaft of the fibula.

Diagnosis. The front portion of the foot is shortened while the heel is more prominent than normal. The lower end of the tibia protrudes over the dorsum of the foot and the sharp edge of its articular surface can be distinctly felt. The extensor tendons and the tendo Achillis are tense and prominent. It may be distinguished from a supramalleolar fracture by the fact that the malleoli in the latter have moved backward with the foot, while in a dislocation backward they are prominent at some distance in front of the heel.

Treatment. Reduction is usually effected by forced plantar flexion, the foot being pulled forward and the lower end of the tibia being pushed backward. These steps are then followed by dorsal flexion of the foot.

After reduction, the leg should be immobilized for three weeks in a molded posterior splint. Light passive motion can be begun during the fourth week. In old unreduced cases an arthrotomy is indicated.

Forward Dislocations. These are much rarer than the backward form. They are usually due to a forced dorsal flexion of the foot. This form is less often accompanied by a fracture of the malleoli than is the case in the backward dislocation. The fibula is seldom broken, the usual seat of the fracture being in the tip of the internal malleolus or in the articular surface of the tibia.

Diagnosis. The whole foot appears to be lengthened. The prominence due to the heel has disappeared; the upper articular surface of the astragalus can be felt, the tibia and the malleoli being nearer to the heel.

The condition can be differentiated from a fracture of both bones of the leg above the malleoli by the fact that in a forward dislocation the malleoli are further back than normal, while in a supramalleolar fracture they have moved forward with the foot.

Treatment. Reduction is readily effected by marked dorsal flexion of the foot, pressure being made in a forward direction upon the lower end of the tibia, and the foot pushed backward. Plantar flexion now completes the reduction. The after treatment is the same as in the backward form.

Lateral Dislocations. The other forms of dislocations seen in the ankle are those in a lateral direction, either inward or outward. The diagnosis is usually easy. The upper convex surface of the astragalus is directed toward the external malleolus and can be felt there. The inner border of the foot is raised; the outer rests upon the bed.

This form of dislocation is very frequently a compound one, or it is accompanied by fractures of the bones of the leg or of the astragalus; but it may occur without these injuries.

Treatment. The treatment of these lateral dislocations differs but little from that of fractures of the lower end of the tibia and fibula. Reduction is effected by adduction or abduction of the foot. The chief danger is from infection on account of the extensive injury of the skin and soft parts. If reduction is impossible, perform an arthrotomy.

Subastragaloid Dislocation. Two forms of dislocation can occur in the joint between the astragalus and the two tarsal bones (os calcis and scaphoid) with which it articulates. In the true subastragaloid form, the astragalus continues to articulate with the tibia and fibula, but it is displaced from its articulation with the os calcis and scaphoid. In the second form of subastragaloid dislocation, the astragalus is completely separated from its articulation with the bones of the leg as well as with the calcaneus and scaphoid. To this form the name total dislocation of the astragalus is given.

True Subastragaloid Dislocations. These dislocations may occur in four directions, inward, outward, forward, and backward.

Dislocation inward. The most frequent cause is a forcible adduction of the foot combined with violence acting in the direction of the long axis of the foot. The diagnosis can be made from the position of the foot. The foot is adducted and rotated inward, as in a case of clubfoot. The sole of the foot is directed inward. The inner edge of the foot is concave and shortened while the outer edge appears lengthened. The external malleolus and head of the astragalus are very prominent on the outer side of the foot. Below and behind the inner malleolus the scaphoid projects beneath the skin.

Dislocation Outward. This occurs after forced adduction of the foot. The symptoms are the opposite of those of the inward variety. The foot is in the position of a flat foot, its inner edge depressed and outer edge raised. The inner malleolus is close to the sole of the foot, and in front of it the head of the astragalus forms a prominence. The injury is not infrequently compound, so that the astragalus presents into the wound.

Dislocation Backward. The cause is usually a plantar flexion of the foot. The signs are very pronounced; the head of the astragalus can be seen and felt lying upon the upper surface of the scaphoid and cuneiform bones. The anterior portion of the foot is shortened while the heel is lengthened and the tendo Achillis is very prominent.

Dislocation Forward. This follows forced dorsal flexion of the foot, the patient falling forward after landing with his heels upon the ground. The diagnosis can be made because of the lengthened anterior portion of the foot and the shortened heel. An important point in the diagnosis of subastragaloid dislocation is the absence of any prominence due to the projection of the body of the astragalus, in front, behind, or to either side of the malleoli, as is seen in the case of the tibiotarsal dislocations. A second diagnostic point is the abnormal position of the calcaneus and scaphoid with relation to the malleoli and astragalus. The swelling is usually so great that a diagnosis is very difficult without the use of the X-ray.

Treatment of Subastragaloid Dislocations. Reduction can usually be effected in recent cases by manipulation and traction. In the inward variety the existing adduction is at first increased. Pressure is now made over the outer side of the adduction and the inner side of the foot, and the foot is then strongly abducted. In the outward variety, the abduction is first increased. Pressure is then made over the outer side of the foot until reduction is effected. In the backward variety, the plantar flexion is first increased and the foot is then strongly flexed in the opposite direction. In the forward type, forced dorsal flexion will effect reduction. The foot should be placed upon a posterior molded splint for three weeks, after which passive motions are begun. If the reduction is impossible, an arthrotomy with excision of the astragalus may be necessary.

Total Dislocation of the Astragalus. This form of dislocation is much more frequent than those of the ankle joint proper, or of the articulation between the astragalus, calcaneus, and scaphoid. The displacement of the astragalus may occur in one of six directions: forward; outward and forward; inward and forward; inward; backward, and by rotation.

The most frequent variety is the "outward and forward." In this variety the foot is rotated markedly inward and the external malleolus is very prominent. The foot is in a clubfoot position. The dislocated astragalus can be felt as an irregular angular bone just below the external malleolus.

Treatment is the same as in subastragaloid dislocations.

Dislocation of the Metatarsal Bones. This may be either complete or incomplete at Lisfranc's joint. It occurs most often in an upward direction. The dorsum of the foot is more convex than normal, while the sole of the foot is flattened. One can see and feel the displaced ends (upper) of the metatarsals on the dorsum of the foot. The foot is shortened and the toes point inward.

Dislocations of the individual metatarsal bones are much rarer. The middle ones are displaced upward, and the first and fifth, inward and outward respectively.

Dislocation of the Toes. This occurs most often in the metatarsophalangeal joint of the great toe after forcible flexion. The dislocation may be complete or incomplete. In the former case, the proximal end of the first phalanx and the dorsum of the foot are prominent, and the head of the metatarsal bone projects on the sole of the foot. The reduction of toe dislocations presents no difficulties.

SPRAINS

Definition. A sprain is a joint wrench due to a sudden twist or traction, the ligaments being pulled upon or lacerated and the surrounding parts being more or less damaged.

Sprains of the Ankle. On account of its flexibility and constant use in weight-bearing, the ankle is the joint most frequently sprained.

Sprains are common in a limb with weak muscles; in a deformed extremity in which the muscles act in unnatural lines, and in a joint with relaxed ligaments.

A joint, once sprained, is very liable to a repetition of the damage from slight force.

Symptoms. The symptoms manifested in a sprain are as follows: severe pain in the joint; nausea and sometimes syncope; impairment, or loss of motion; severe pain upon motion; early swelling if hemorrhage is severe—in any case swelling begins in a few hours; movement of the joint becomes difficult or impossible; the tear in the ligament may be distinctly felt; in a day or two pain and tenderness become intense and discoloration becomes marked.

Diagnosis. Usually the diagnosis is easy to make, but in all doubtful cases an X-ray picture should be taken in order to be certain that a fracture does not exist.

Treatment. The first indication is to arrest hemorrhage and to limit inflammation. For the first few hours apply pressure and an ice-bag. Wrap the joint in absorbent cotton, wet with iced water; apply a wet gauze bandage, and put on an ice bag.

In a mild sprain, use lead and opium wash. In a severe sprain, place the extremity upon a splint and apply to the joint flannel kept wet with lead-water and laudanum, iced water, tincture of arnica or alcohol and water. If the pain is severe, a small dose of morphine should be given.

Judicious bandaging limits the swelling. When the acute symptoms begin to subside, rub stimulating liniments, such as chloroform or arnica, upon the joint once or twice a day and employ firm compression by means of a bandage of flannel or rubber. Later in the case use hot and cold douches, massage, passive motion and the bandage.

Another method of treatment of sprains of the ankle is by strapping with adhesive plaster, but it is advisable only for slight injuries. In severe cases, in which extensive laceration of the ligaments is suspected from the marked extravasation, it is best to immobilize the foot in a plaster-of-Paris splint for two weeks; later baking in a hot-air oven (see "Arterial Hyperemia") with massage, and active and passive motion are advisable.

In simple sprains, the fixation does not produce serious stiffness, and without fixation the repair of the ligaments is only partial. In the latter case, the result is weakness of the ligaments and an instability of the foot which leads to frequent recurrence. This explains many habitual sprains. On the other hand, under appropriate treatment, a sprain should recover without leaving any functional disturbance.

CHAPTER XVII

DEFORMITIES

PES PLANUS, OR FLAT FOOT

The terms *weak foot* and *flat foot* will be used to designate the *mild* and the *severe* forms of the same condition which include all the deviations from the normal height of the arch of the foot.

Flat Foot may be congenital or acquired, the former being a very infrequent deformity, and the latter one of the most common pathologic conditions.

Congenital Flat Foot is a deformity of infrequent occurrence, and in some cases is associated with defective formation of the bones of the foot. In this condition the whole foot is displaced outward in relation to the leg; the sole is rolled outward, the inner malleolus is prominent and the foot is abducted on itself, and in severe cases, it cannot be replaced in its normal position on account of the contracted tissues.

Treatment. The foot should be massaged and, by gentle manipulation, forced into its proper position and held by a plaster-of-Paris dressing, changed at the proper intervals. A tenotomy may be required to bring the foot into its proper position.

When the child begins to walk, a well-fitting arch support should be worn.

Acquired Flat Foot. The common form of acquired flat foot is the static variety, which is an expression of a disproportion between the body weight and the sustaining power of the muscles and ligaments.

Common Causes. 1. The use of improper shoes is by all means the most frequent cause of flat foot, and frequently makes all of the following causes more pronounced.

2. Weakness and insufficiency of the muscles, resulting from poor general condition; advancing age; convalescence from acute illness; from childbirth; and from injuries of the leg, especially fractures.

3. Prolonged standing, especially on hard wood and stone floors.

4. Rapid body growth.

5. Rapid increase in body weight.

6. Excessive weight bearing.

7. Shortened condition of the gastrocnemius muscle.

Other causes are rickets; inflammation of the ankle joint, as in tuberculosis; or, as a result of a badly treated fracture of the ankle-joint; or, as a result of paralysis of the muscles of the inner side of the leg.

Pathology of Acquired Flat Foot. The pathologic condition is due to change in the relations of the bones rather than to any change in the bones themselves. The abnormal position is an exaggeration of the normal yielding of the foot under weight bearing. The front of the astragalus rotates inward, and with it the bones of the leg turn at the hip-joint.

The deformity is essentially a displacement of the astragalus on the bones of the tarsus. The scaphoid, cuneiform, and the base of the first metatarsal move downward and inward with the head of the astragalus; the outer border of the foot is made more concave and the inner border becomes convex in extreme cases. In the severest cases, the head of the astragalus, and scaphoid may be displaced below the plane of the other bones. The ligaments are respectively shortened and stretched in the severest cases and there is a loss of motion in certain of the tarsal articulations, due to faulty apposition of joint surfaces, and to constant strain.

Symptoms. The feet burn and tire easily and feel stiff and lame. They may swell, and the size of the shoe worn must be then increased. Later, a painful period generally begins in which walking is avoided and a dragging pain in the arch and behind the inner malleolus is noticed. This is increased by walking and standing and tender points may be found under the scaphoid and on the upper surface of the heel. The foot feels strained and irritated and is a constant source of discomfort. The inner malleolus is generally more prominent and the foot is displaced outward in relation to the leg. The height of the arch is somewhat diminished; it may be much lowered, or it may be flat on the ground.

When the foot is really flattened, it presents two types, one the *flexible flat foot*, in which the arch can be restored by gentle manipulation; the other, the *rigid foot*, which is held by structural changes in the position of deformity.

An intermediate type is sometimes seen, in which the peroneal spasm is so great that the foot is held abducted and everted as long as the spasm lasts (spastic flat foot.)

Some symptoms of flat foot that are less generally recognized, which are of great value in diagnosis are: corns, ingrowing nails, callosities on the sole of the front of the foot, enlargement of the great-toe joint, and pain (especially at night) in the calves of the legs and backbone, which is aggravated by standing and walking.

Diagnosis. The diagnosis of flat foot, whether flexible or rigid, is made chiefly by inspection. The difficulty comes in the milder cases, which form the bulk of those seen, and in which the changes in form are slight.

Symptoms. The symptoms, as described by the patient, are the most reliable and points of tenderness under the arch or heel would help to confirm the diagnosis. Some help may be obtained from a wet impression of the foot, on a piece of paper, but the slighter cases show but little changes in the imprint. In most normal feet, the outer border of the foot touches the paper, and in flat foot, only two areas bear the weight, one on the inner side of the front of the foot, and one under the inner part of the heel. An X-ray picture is often of great assistance.

The diagnosis of rheumatism is frequently made in flat foot, and is often the source of much misdirected treatment. Rheumatism should be diagnosed only in connection with unmistakable symptoms of rheumatism in the upper extremities.

So-called "rheumatic" pains in the knees and hips may be secondary to flat foot.

Prognosis. As a rule, this condition does not recover spontaneously. Under ordinary conditions, uncomplicated cases should be at once relieved by proper treatment, and in time should be cured.

Unfavorable factors are: great weight; disease of the ankle-joint; the presence of bony spurs under the os calcis.

The prognosis is more favorable in young adults than in persons of advanced age. Patients, who without relief have worn the ordinary supports sold at the stores will, as a rule, manifest extreme sensitiveness as to the fit of any of the supports which may be applied.

Treatment. The foot must be restored and held in its normal position and measures must be adopted to quiet local irritability or inflammation, and to strengthen the muscles. The best treatment does not consist in the permanent wearing of a flat-foot support; the support should be regarded in the same light as one uses a crutch in a fracture of the leg.

As a preliminary to all treatment, the use of proper shoes must be insisted upon. A shoe should be as wide in front, as the unshod foot, when bearing the weight of the body.

Supports. Flexible supports may be made of boiler felt; one objection to these is their liability to stretch. They are of service in young children, in mild cases, and in convalescent cases where it is desirable to have the patient use a flexible instead of a stiff support in order to bring the muscles into play.

Rigid supports are best made of tempered spring steel (18 to 20 gage), forged hot to fit a cast of the foot. They may also be made of phosphor-bronz, celluloid or aluminum.

The shape of the plate is largely a matter of judgment. The easiest way to determine the shape of the plate to be used in a given case is to have the patient stand with the operator's hand under the inner side of the foot; the operator then places the foot in the normal position and notes where the pressure must be applied to secure the proper correction; when the anterior part of the foot is flattened, a slight dome must be constructed in the front of the plate; when the os calcis is clearly tilted over, the plate must have two flanges at the heel to hold it in place. In general, the plate must reach forward to a point just behind the great-toe joint, and must furnish support as far as the front of the heel. The plate should be higher on the inner side, and a flange formation is generally necessary to accomplish this. An outer flange prevents the foot from slipping off the outer side of the plate. When the foot no longer requires support, the plate should be gradually discontinued.

The "Thomas" sole may be used in mild cases. This is made by building up the inner part of the sole of the shoe one-eighth to one-quarter of an inch higher than the outer side, thus securing a slight inversion of the foot.

Exercise and massage of the deficient muscles should form a part of the routine treatment in all cases of flexible flat foot.

To diminish local inflammation and irritability, the foot should be soaked in hot water; hot and cold alternate douches should be applied, and hot-air treatment and massage should be employed.

Rigid Flat Foot. Rigid flat foot cannot be successfully treated until the position of the foot is corrected. The patient should be anesthetized, and, by the use of a wedge as a fulcrum, the bones should be forced into position. A pressure of about two hundred pounds is generally necessary to effect this reduction. After this, the foot is placed in a plaster cast, in extreme adduction and is allowed to remain thus encased for three weeks. After this, a properly fitted plate should be worn. The results are usually satisfactory.

Operative Treatment. Cases that have resisted all other forms of treatment, may be cured by the removal of a wedge-shaped piece of bone, with the base downward and inward at the point of greatest inward convexity, that is, in the neighborhood of the head of the astragalus. Osteotomy of the front of the os calcis and neck of the astragalus will at times be necessary for a radical cure.

Many other operative procedures have been advised for flat foot and they have been employed with varying successes.

Hallux Flexus or Hammertoe. The upward prominence of a toe (usually the second or third) in a rigid position, is known as *hallux flexus* or *hammertoe*. In this condition the toe is flexed in its second joint so that the end bears on the ground, while the junction between the phalanges makes a prominence upward. Helomata and callosities may develop on the end of the toe, but the chief discomfort is in the disturbances which arise on the prominence which presses against the side of the foot-gear.

Treatment. A knowledge of the forces at work will show how futile must be any effort to correct this deformity by strapping or bandaging. There is a shortening of the plantar fibres of the lateral ligament of the joint. The trouble does not lie in the flexor tendons, as it seems, and operations directed to this point fail. Even with incision of the lateral ligaments, followed by the application of a splint, recurrences are common and amputation must be the procedure.

The condition described as hammertoe may exist in several or in all of the toes, the great toe being least often involved. This occurs most often as a result of wearing improper shoes, but is sometimes the consequence of paralysis.

Flexed or Clawed Toes. Extreme flexion of all but the great toes causes the weight to be borne by their dorsal aspect. In this condition the toes, and especially the small ones, develop painful helomata on the prominent joints, and the small toe may become the source of great discomfort.

Treatment. Radical surgical measures are here indicated. Tenotomy or amputation is essential to a cure.

Painful Heel. Painful heel is a suggestive but unscientific term applied to tenderness of the under side of the heel. It is associated with one of the following conditions:

1. Spurs running out from the under side of the os calcis found by the aid of the X-ray.

2. Inflammation of the bursae under the os calcis.

3. Flat foot.

4. Gonorrhœa.

5. Focal infection.

Treatment. Where a spur of bone causes the unpleasant symptoms, the excrescence should be excised.

When focal infections are the primary cause of painful heel, operative procedure to remove the source of infection is imperative and will prove curative.

Palliative measures are: massage, douches, hot air, a metal plate worn under the painful area, rest. The back of the foot should be cut away to relieve pressure.

Metatarsalgia—Morton's Disease. Metatarsalgia is characterized by an acute pain, cramplike in character, occurring at the base of the third or fourth toes.

The pain comes on suddenly while the foot is in action, and is usually accompanied by a "snapping of the bones." The pain is so acute that it is not uncommon for the patient to seek relief by taking off the shoe and rubbing the foot.

In persons suffering with this condition it will be regularly noticed that the weight is thrown upon the ball of the foot, on the metatarsophalangeal joints, either because of a weak foot, or because of a tendency of the toes to turn up.

Treatment. 1. Proper strapping to raise the arch and bring the ends of the toes down.

2. A pad across the ball of the foot *behind* the metatarsal heads, also brings the toes down.

3. Recommend shoes, wide across the ball, with a higher or lower heel than ordinary, as the case indicates.

Hallux Valgus. The term *hallux valgus* is applied to a deviation or displacement of the great toe outward, toward the outer border of the foot.

In normal feet, the line of the great toe when prolonged backward, should pass through the centre of the heel. This relation in civilized communities is seen only in the feet of infants. In adults it is observable only in the bare-footed races.

Cause. It is frequently associated with flat foot, gout and rheumatism, but it is primarily due to the use of inappropriate foot-gear. It is only considered pathologic when the deviation is more than fifteen degrees.

Pathology. The displacement outward (which reaches 30 to 40 degrees in the average case and may reach 90 degrees) of the phalangeal part of the great-toe joint, uncovers the inner part of the head of the metartarsal bone, and here the cartilage degenerates, and the bone becomes condensed at its outer part. The inner lateral ligament is lengthened and thickened and the sesamoid bones become displaced outward and are often thickened.

Under the skin, at the inner and prominent aspect of the foot, is to be found a bursa, which is liable to inflammation under pressure, and is known as a bunion. The inflammation in this sac may extend to the joint and thus disintegrate it.

Symptoms. The toe is displaced outward and a reddened and shiny condition of the thickened skin exists over the inner prominence and perhaps over the top of the toe joint. The great toe if seriously displaced, must lie over or under the other toes, the former being the more common position. In other cases the second toe may be crowded up as a hammertoe. The joint is painful and the inner toes, being crowded to the outer side of the foot, are the seat of corns and callosities. Flat foot is frequently associated with this condition.

Treatment. In mild cases, the stocking should be split to allow a separate stall for the great toe, and broad toed boots should be worn. If flat foot exists, a support should be supplied for its aid in restoring the position of the great toe. In severe cases, nothing short of an operation is likely to be of value. A toe-post may be worn for a time in mild cases.

Amputation of the head of the metatarsal bone gives uniformly good results.

The toe is straightened and flexible; ankylosis with this operation does not occur.

In operations for hallux valgus there are two distinct purposes acting as determining factors in making a choice in a given case as to which is indicated. These are: (1)the radical operation for the correction of the deformity, and (2)the palliative operation for the alleviation of symptoms by the removal of the hypertrophied portion of the metatarsal head which is exposed to pressure. Among operations in the first mentioned class, the one known as the Mayo operation is, in all probability, the best. The entire head of the metatarsal is amputated, and the bursa is turned in over the cut end of bone, to diminish the amount of shortening and to prevent ankylosis of the joint. This latter consideration, however, is an unnecessary one, for in operations within this joint, ankylosis does not occur when the synovial surface of the phalanx is left undisturbed, even when the bursa is not employed as an intervening pad.

In the other class of operations for the relief of symptoms, no attempt is made to straighten the toe. A wedge-shaped piece of the exostosis is removed, against which pressure has caused symptoms.

A palliative operation devised by Dr. Robert T. Morris of this city, is one easy of accomplishment and serves every purpose where a radical operation is interdicted. It is known as the "button-hole" operation because of the fact that only a small incision is made immediately above the protuberant bone through which a sharp chisel is inserted, cutting off the offending "button" of bone.

An operation which in the hands of the authors has proven of distinct value, and which has probably not been previously described eliminates both the deformity and its painful symptoms. This operation which is described below, is less severe than other radical operations and not very much more so than the usual palliative ones.

The incision is made on the dorsum of the great toe over the offending joint and just to the inner side of the extensor tendon. This tendon is held to the outer side, out of the way. The knife penetrates the capsule of the joint and opens it above and laterally.

An effort is made to preserve the integrity of the capsule below (floor) as *only the intra capsular end of the metatarsal is removed*. These two factors are of the utmost importance. When the joint capsule is slit open along its dorsal and two lateral aspects, sufficient room is obtained for the insertion of the wire saw, and all of that portion of the metatarsal lying within the joint proper is removed. There is thus accomplished a correction of the deformity with very little shortening of the great toe. Usually its length after this operation is about the same as the second toe.

The next step in the operation is closure of the synovial sac or joint capsule. A stitch on either side and two above are all that is necessary. The floor of the sac remains intact and nothing beneath it, in the ball of the foot, has been disturbed. Many operators invade this area and remove the sesamoids. This is unwarranted as the transverse level of the ball of the foot is lost, and the weight is put directly upon the newly formed joint, depriving it of its normal support, or of padding from below.

One other omission in this operation is that of the bursal flap over the raw end. This is found entirely unnecessary as results prove, and its omission hastens healing considerably. The bursa over the metatarsophalangeal articulation in these cases is nearly always inflamed, and consists of a mere fibrous pad. Its dissection from the normal position is a real loss at that site, and of questionable benefit over the cut bone, as motion in the joint is as good or better without it.

The skin closure is made without drainage, and *no wet dressing employed* for fear of the solution filling the cavity whence the bone was removed and carrying with it infectious material. A dry sterile dressing is all that is required, and a splint to maintain a straight position for the toe.

Four or five days complete rest for the part are ordinarily sufficient. Following this, walking about the room is permitted with the aid of a stick. After ten days, when the patient can get about fairly well without the assistance of a stick, the foot may safely be shod with an "arctic" of sufficient size.

CLUBFOOT OR TALIPES

The most common form of clubfoot, and therefore the deformity of that character most frequently encountered, is characterized by inversion of the sole of the foot, elevation of the heel, and a twisting and turning of the front part of the foot. This deformity is typical of *congenital* clubfoot, which, as stated, is the most common form of that deformity. The *acquired* form is usually the result of infantile paralysis.

Congenital Clubfoot is most frequently double, and males are more frequently affected than females; in unilateral or one-sided clubfoot, one side is not more frequently affected than the other.

Etiology. Very little is known as to the cause of congenital clubfoot but it is not infrequently associated with other congenital deformities. It appears to be hereditary in a great many instances. The greater number of cases appear without definable cause, except perhaps from intra-uterine pressure. There are, however, a number of these cases that are associated with malformation of the bones of the foot and leg, such as absence of the scaphoid; defect of the tibia; fusion of a number of the tarsal bones.

Pathology. The sharp adduction and plantar flexion, at the tarsal joints, produce a deformed position of the foot. As a result of these, the heel is small and elevated; the dorsum of the foot is prominent; and the outer border usually, and, in extreme cases, the dorsum of the foot, bears the weight of the body in walking and in standing; the sole of the foot is bent sharply in, and twisted at the tarsal joint. In fact, all the bones are changed in shape, and the inner muscles, tendons and ligaments are shortened by contraction, while the ones to the outer side are lengthened.

The distortion of certain individual bones is of importance. The astragalus is the seat of the most important changes. It is tipped downward at its front end, and its posterior part articulates with the tibia, its anterior articular surface projecting under the skin; its neck is elongated and bent inward and downward, so that its scaphoid articulation faces inward and downward and not forward.

This is the most important change in clubfoot, because the anterior end of the astragalus, the head of the bone, carries inward and downward with it the scaphoid, the three cuneiforms, and the inner three metatarsal bones. The scaphoid articulates with the inner side rather than the front of the astragalus and, in extreme cases, forms a joint surface with the inner malleolus. It may be somewhat changed in shape, being flattened and drawn inward and upward.

The os calcis is generally poorly developed, and its front end is rotated downward, and bent inward; the outer surface of the bone is more convex and the inner surface more concave than normal, and since the anterior facet looks inward and downward, it carries with it the cuboid and the two external metatarsal bones. The changes in the other bones are not important; the chief obstacles to reduction lie in the os calcis and in the astragalus.

Soft Parts. The muscles, ligaments, tendons, and fascia at the lower and inner side of the foot are shortened, and lengthened at the outer and upper side. The plantar fascia being one of the chief obstacles to reduction, the tendons are displaced, especially those on the inner side of the foot.

Symptoms. Double clubfoot is usually accompanied by an awkward and unsteady gait, in which each foot is in turn lifted high to clear the foot on the ground, and the *toeing in* is, of course, excessive. The weight is borne on the outer side of the foot, and all elasticity of gait is absent.

On the outer border of the foot, where the weight is borne, callosities and bursae develop; the calves of the legs are small, and the knee joint may be lax.

The gait in single clubfoot is less awkward, but characterized by the same features. The foot is rigid in the deformed position, and in cases of marked deformity, the foot cannot be manipulated into the normal position.

Diagnosis. Congenital clubfoot cannot be mistaken for any other condition. The diagnosis is self-evident.

Prognosis. There is no tendency of this deformity to right itself, or to improve. Early and proper treatment will, if continued long enough, insure a cure in children and an improvement in adult cases; but it must be remembered that there is a decided tendency to relapse, even after operation, unless the foot is kept in an overcorrected position for a number of years.

Treatment. In young infants, treatment should be begun as early as two weeks after birth and should consist in frequent gentle massage and manipulations. After the part can be brought into an overcorrected position by gentle manipulation, it should be put up in a plaster cast, for a period of three weeks and this treatment should be continued until the position of the foot is corrected.

The manipulations consist in grasping the dorsum of the foot gently but firmly with one hand, and holding the leg with the other. The foot is then dorsally flexed and everted. This treatment should be repeated at least three times a day and should not be rough enough to cause the infant to cry.

Treatment of clubfoot in older children and adults is a much more difficult proposition and consists in the combination of two or more methods of procedure.

In order to correct the extreme adduction in these cases, extreme force must sometimes be employed. This may be accomplished by bending and bearing down on the foot, with its outer border resting on the apex of a wooden wedge. The rotation of the foot is corrected by grasping the foot in one hand, and the heel in the other, and twisting with the necessary amount of force. The inversion of the sole is also corrected by the use of this wedge as a fulcrum.

In this way the tendo Achillis and the plantar fascia are stretched, and the dorsal flexion is secured by laying the patient on the face with the knee bent and the front of the thigh resting on the table. The lower leg is then vertical, and by bearing down on the front of the foot with the necessary amount of force, dorsal flexion of the foot is secured, and by hooking the fingers around the os calcis, its position is improved.

A modified Thomas wrench may be used in the correction of clubfoot; but this must be done with great care, as the violence practised in this method, the tearing of the ligaments and other soft parts, is often attended with great danger; osteomyelitis, tuberculosis, neuritis, and even death from fat embolism, and extensive sloughing of the soft parts are not infrequently seen after the use of this and other bone crushing instruments.

The removal of a wedge of bone from the outer side of the foot and the removal of the neck of the astragalus are employed. Tenotomy and the transplantation of tendons are also often practised, when other methods of treatment fail.

Acquired Clubfoot. The cause of acquired clubfoot maybe infantile paralysis, joint disease, traumatism, or it may be due to affections of the brain or spinal cord.

Paralysis. Infantile paralysis affecting the muscles of the front and outer side of the lower leg, will result in a condition similar to congenital clubfoot. Other paralytic causes are: spastic or cerebral paralysis, hereditary ataxia, etc.

Traumatic. A condition resembling clubfoot may result from improperly treated fractures of the ankle-joint or tarsal bones.

Joint Disease. In tuberculosis, arthritis deformans, and other diseases of the ankle-joint, a condition similar to clubfoot is sometimes seen as a result of muscular contraction.

Talipes Equinus is rarely congenital. It is usually due to infantile paralysis of the extensor muscles, or to cicatrical contraction of the calf muscles, as a complication of hip disease. It varies from inability to flex the ankle beyond a right angle, to walking on the heads of the metatarsal bones. The astragalus is partially displaced forward and forms a prominence on the dorsum of the foot; the plantar fascia is shortened and callosities and bursae are formed under the heads of the metatarsal bones. Primarily, the obstacle to reduction is the tense

Achilles tendon, and in advanced cases the shortened plantar fascia and posterior ligament of the ankle-joint constitute obstacles.

Talipes Equino-Varis (down and in foot) is the most common form of this deformity.

It is either congenital or acquired, and in the latter case it is due to infantile paralysis of the extensor and peroneal muscles. The heel is drawn up, and the anterior half of the font is drawn inwards and inverted. The inner border of the foot is shortened, and in neglected cases the patient walks on the outer side of the cuboid, under which a bursa is formed. Secondary contraction of the plantar fascia, ligaments, and short plantar muscles follows. There is a great increase in the obliquity of the neck of the astragalus in congenital cases, so that the scaphoid and anterior half of the foot, together with the dorsal tendons are carried inward. As a result of the equinus, the upper surface of the astragalus projects forward, and only its posterior portion comes in contact with the tibia and fibula. The ligaments of the inner side of the foot are shortened and the shape of the other tarsal bones is secondarily altered.

Talipes Equino-Valgus (down and out foot). This condition is rare as a congenital deformity. The anterior half of the foot is deflected outward, and the inner border comes in contact with the ground. The scaphoid is placed outward, and the head of the astragalus projects into the sole.

The acquired variety results from paralysis of the tibialis posticus and flexors, with secondary contraction of the peronei muscles.

Talipes Calcaneus is rare as a congenital deformity. It is usually the result of infantile paralysis of the muscles of the calf. The patient walks on the heel, and the anterior half of the foot is drawn up. Valgus or varus are associated with it; the more common form is talipes calcaneo-valgus.

Talipes Cavus (Pes Cavus), or hollow foot, is a condition in which the arch of the foot is greatly exaggerated. It is rarely congenital but is frequently seen in connection with clubfoot, especially in its paralytic forms. In its mildest form, it exists in a highly arched foot, often hereditary. It may also be the result of too short shoes (Chinese ladies' foot).

Treatment. The condition is best remedied by division of the contracted soft parts, a forcible reduction of the bones, held in place by plaster of Paris. When the patient begins to walk, it is advisable to have a stiff, flat, steel plate placed in the length of the shoe between the layers of the leather sole, running from which, over the dorsum of the foot, is a stout leather strap. At each step, downward pressure is thus exerted on the dorsum of the foot.

CHAPTER XVIII

THERAPEUTIC MEASURES

HYPEREMIA

Hyperemia as a therapeutic agent was described by Bier and is of two kinds, *active* and *passive*. The former is the same as the *arterial*, while the latter is the *venous*. Between the blood of active and passive hyperemia there are important physical and chemical differences, the one containing much free oxygen with but little carbonic acid and alkali, while the other presents the exactly opposite character.

In active hyperemia normal elements of the blood are kept in active motion, while in the passive form they are allowed to escape, more or less, into the tissues.

Hyperemia possesses a great many properties:

1. Power to diminish pain.

2. Bactericidal action.

3. Absorptive property.

4. Solvent action.

5. Nutritive power.

6. Suppression of the infection.

Hyperemia may be produced in three ways; *first*, by means of the elastic bandage or band; *second*, by cupping glasses, and *third*, by hot air. The first two produce venous or passive hyperemia, and the third, arterial or active hyperemia.

Passive Hyperemia. This obstructive hyperemia is produced by means of a thin, soft rubber elastic bandage, two or three inches in width, better known as the Esmarch, or Martin bandage. When this is applied moderately tight around a limb about six or eight turns, one layer overlapping the other, pressure is evenly distributed over a comparatively wide area, causing the subcutaneous veins below the constriction to swell; the extremity becomes somewhat bluish red in color, also larger and edematous, giving a feeling of warmth to the touch.

The rubber bandage, properly applied, should not cause any uncomfortable feeling and there should be absolutely no pain present. At all times one must be able to feel the pulse below the site of the bandage. If the bandage is applied too tight, the skin of the limb looks grayish-blue and there appear whitish, or vermilion colored spots, which grow larger and larger, as long as the too tightly drawn bandage is on. Paresthesia and pain, with disappearance of the pulse, can also be noted.

The two cardinal rules to be observed in the application of the bandage are: (1) absolutely no pain with the application of the bandage; (2) the pulse at all times must be felt below the bandage.

In cases which require the bandage to remain in place from sixteen to twenty hours each day, it will be necessary to first apply a soft flannel bandage underneath the rubber one in order to prevent pressure necrosis.

Frequently changing the location of the bandage up and down the extremity, and treating the skin with alcohol rubs, will also be helpful to patients with a tender skin. The elastic bandage must always be placed upon a healthy area, proximal to the diseased part. All dressings should be removed while the compressing bandage is on, in order that the part may become hyperemic.

Wounds or sinuses are covered with sterile gauze and kept in place with a towel, fastened with a few safety pins.

In acute inflammation, septic wounds and phlegmons, the increased inflammation is apt to frighten the beginner, but this is a desired phase of the treatment.

As a prophylatic against infection, it cleanses the wound, produces a local immunization and reaction before the infection has a chance to work; the earlier the bandage is applied the more remarkable is the effect.

For incised wounds of the foot with division of the muscles and tendons, if the tissues are not too seriously injured, the muscles and tendons should be united and the skin closed with interrupted sutures sufficiently far apart to allow free excretion. No drainage is employed and a slight compressing dressing is applied. The elastic bandage is applied very lightly, producing only a slight venous engorgement and the bandage should remain on from ten to eighteen hours a day.

As soon us the symptoms of acute inflammation subside, the time of application of the bandage is reduced. If signs of suppuration are present, the wound should be promptly opened and the pus evacuated. The knife takes care of the pus; hyperemic treatment fights the infection.

In gonorrhoeal arthritis of acute or chronic nature, and in cases of tuberculosis of the bones and joints, the passive form of hyperemia is especially indicated.

The use of cupping glasses is limited to abscesses, furuncles and sinuses.

Active Hyperemia, or arterial hyperemia, is produced by means of hot-air boxes such as the Tyrnauer electric apparatus, or the gas apparatus of Betz.

Active hyperemia increases the arterial blood to any part of the body, thus favoring the absorption of chronic exudates, infiltrates, adhesions, etc. Dry, hot air permits the use of a high degree of temperature without injury or pain to the respective part.

For neuritis of the foot, ulcers, especially diabetic, perforating and varicose, and for the stiffness following a chronic inflammation, or after a fracture, the arterial form of hyperemia gives good results.

COLD

Cold, or the rapid abstraction of heat, is a remedial measure that is nearly always available and is possessed of very great power for good in selected cases.

When cold is applied for its limited and local action, it is always used with two objects in view, namely, (1) to cause localized contraction of the blood vessels, which through inflammation are engorged, so that the parts are swollen and reddened; or (2) temporarily to anesthetize or benumb the nerve terminals, for the immediate relief of pain, in the hope that the temporary paralysis may ultimately result in such changes as to produce a cure.

Cold, in some form, is a popular remedy for a sprain, or any injury likely to be followed by inflammatory processes. A very useful remedy for the sprain of an ankle, when it is a recent accident, is to let the patient sit with the foot elevated, with a cloth wrung out in ice water, and an ice bag applied over the affected part.

In the treatment of localized pain or inflammation, cold is used in a number of ways, largely depending upon the will of the physician and the means of the patient. The simplest, cheapest, and perhaps the best method of using cold, is to place cracked ice in a rubber bag, the latter to be thoroughly watertight, lay it over the inflamed part, surrounding it with a towel so as to prevent the moisture, which appears on the surface from condensation, from wetting the clothing.

HEAT

Heat is used locally for a number of purposes in the same manner as cold, and the choice of heat or cold in the treatment of any acute form of inflammation depends almost entirely upon the wish of the patient, who generally can tell at once which will give him the greater comfort.

In sprains of the ankle, nothing compares to a hot foot-bath prolonged for hours, the object being to decrease the pain and swelling, thereby regaining the use of the limb.

The high degree of heat which can be borne by gradually increasing the temperature of the water by the addition of small quantities of scalding water, is extraordinary, and the favorable results obtained are in direct ratio to the height of the temperature.

Between these soakings, the part should be dressed with lead and opium wash, and rubbed with ichthyol ointment or camphor liniment.

Hot-water bottles or bags are also used locally for the relief of congestion and pain.

THE HIGH FREQUENCY CURRENT, OR VIOLET RAY

The Violet Ray or High Frequency Current is one which is in a rapid state of to-and-fro vibration and is applied through vacuum glass attachments or electrodes, which are excited to a beautiful violet color. The discharge may appear to the eye to be a single spark, but it is made up of a number of successive sparks, following each other with such extreme rapidity that they are said to oscillate (change directions) millions of times per second, a speed that the eye cannot note. The rapid oscillations have the effect of producing the following phenomena:

1. the high frequency current is unipolar, that is, does not require a complete circuit.

2. glass does not insulate the high frequency current as it does ordinary electricity.

3. the high frequency current generates enormous quantities of ozone during its flow.

4. the current does not produce any pain.

5. the high frequency current produces a cellular massage.

The contractile effect is expended upon the individual cells making up the tissues, instead of on individual muscles.

If a sedative effect is desired, keep the electrode in contact with the part; if a stimulating effect is desired, hold the electrode away from the surface; the farther away, the longer the spark.

A uniform spark of any length can be produced by administering the current through layers of toweling, or through the clothing; the length of the spark depends upon the thickness of the layers.

The use of the high frequency current in surgery is limited to sprains, stiff joints, neuritic pains, and adhesions due to inflammatory exudates. Fulguration for the destruction of growths is obtained by employing a pointed metal electrode.

RUBEFACIENTS

Rubefacients. These are agents which revulse by causing congestion of the skin:

1. Turpentine. A few teaspoonfuls of oil of turpentine sprinkled over a piece of flannel wrung out of hot water, applied to the skin and covered with oiled silk or dry flannel, constitutes the turpentine stupe. Twenty minutes is the maximum for this application.

2. Mustard. Mustard flour (the black being the stronger), mixed with tepid water into a paste, spread thinly on a piece of muslin or paper, and covered with gauze or thin cambric, is an excellent counterirritant. Few skins will bear pure black mustard for more than ten minutes. Mustard, diluted one-half with wheat or corn flour, and allowed to stand for twenty minutes, should be the maximum strength for application, because blistering must be avoided, that produced by mustard being specially painful. After removing a mustard plaster, greased lint should be applied.

3. Mustard Foot-Bath. A mustard foot-bath consists of one or two tablespoonfuls of pure mustard in a bucket two-thirds full of water at 105°F; the feet may be kept in this for about twenty minutes, a blanket being thrown around the limbs, and including the bucket, to retain the heat.

Revulsives must be used with caution in cases of shock or coma, lest impaired vitality or sensation to pain result in extensive sloughing of the skin.

CAUTERIES

The Actual Cautery is used in the form of variously shaped irons, hatchet-edged, round, or olivary, fitted into wooden handles, and heated in a charcoal furnace.

As a counterirritant, the iron should be heated only to a dull red heat, and should be quickly drawn in parallel lines, about one inch apart, over the skin, avoiding all bony prominences. Compresses wet with cold water, or with some antiseptic lotion, may then be applied.

The Paquellin Thermo-Cautery is a convenient form. It consists of hollow platinum cauteries and a handle covered with wood; a benzole reservoir; a pair of rubber bulbs, like those for a hand-spray apparatus, connected by a tube with the reservoir; a long rubber tube to connect the cautery handle also with the reservoir; and a spirit-lamp with attached blow-pipe.

Screwing on the desired point, the tube from the reservoir is slipped over the handle; the point is heated in the lamp; is removed from the flame; and, compressing the bulbs, which should previously have been connected with the reservoir, benzole vapor is forced into the point, which will heat up, and can be maintained at any temperature by the rapidity with which the bulb is worked. If the point will not heat with the simple flame, attach the bulbs to the blow-pipe on the lamp, and, compressing them, heat the cautery to a bright-red heat, and then connect with the reservoir and proceed as before directed.

Galvano-Cautery. This requires a battery of a few large elements closely coupled, and various curets, knives, and ecraseurs fitting into insulated handles. The chief advantage of this form of cautery is the possibility of placing the instrument in position while cold, and then heating it.

Where hemorrhage is undesirable, a dull-red heat should be maintained, for at a white heat the tissues are divided as if with a knife, and bleeding follows. When the ecraseur is used, needles must be passed at right angles through the healthy tissues, the platinum wire placed behind these, and the wire, at a dull-red heat, slowly tightened.

ELECTRICITY

Electricity. This is used in the form of the *induced current* (Faradism) to exercise and improve the nutrition of muscles, and in the form of the *constant current* (galvanism) along the course of nerve-trunks, to excite their conducting power, or to act as a sedative in neuralgias.

The same current is used to induce chemical decomposition (*electrolysis*) or to cauterize and destroy tissue by heating an encircling wire or by a galvanic knife. Franklinic, or static electricity, is also occasionally used.

Electrolysis. For electrolysis a galvanic battery of thirty or more medium-sized cells is required, with needle electrodes insulated, except near their points.

To destroy a verruca, introduce into it two needles, a short distance apart, each connected with a pole of the battery; then, commencing with a weak current, this must be cautiously increased, the sitting lasting from a half hour to one hour, after which the needles are to be removed and the punctures sealed by collodion.

MASSAGE

Massage. This is employed to stimulate the circulation in the part mechanically; to loosen tissues bound down by adhesions; to diffuse inflammatory exudates over a wider area, thus favoring their absorption; and to change the rate of the circulation to a point compatible with rapid absorption and normal nutrition.

Four distinct varieties of manipulation are found to be most generally useful:

1. rubbing, or stroking

2. kneading

3. tapping, or percussion.

4. passive and active moments.

Stroking consists in gentle rubbing directed from the periphery upward, commencing the process above the inflamed part and continuing it over the diseased area; the pressure, at first light but finally firmer, will force the exudates into the tissues above, which have been emptied by the preparatory rubbing.

Kneading means rubbing the part circularly with the pulps of the fingers and the thumb or the palm of the hand, and is best combined with pinching up of the skin or muscles singly or together, and gently rolling them between the fingers and palms.

Percussion is effected by tapping the surface over the diseased part with the tips of all the fingers held on a level, or with the ulnar side of the hands, or, after covering the part with a towel, three parallel pieces of stiff rubber tubing, fixed in a handle (a muscle beater), may be employed, gently striking the part transversely to its long axis.

Passive movements should be made at the close of each sitting if a joint is concerned.

Massage is sometimes advisable twice daily, but often once a day or every other day is better; each sitting may last from fifteen minutes to one hour.

EXAMINATION BY RADIOGRAPHY

X-Ray Examination. This method of examination depends on the property of penetration of matter possessed by a radiation from an electrically excited Crookes' tube. This radiation has been proved to lie outside the spectrum, and has been named X-ray.

It may, for purposes other than those required by the expert, be looked upon as a source of light which has the property of penetrating the tissues to a greater or less extent according to their density, and the shadows cast by it can be recorded on a photographic plate, or may be viewed with the naked eye by means of a screen composed of a thin layer of barium platinocyanide, a substance which becomes highly fluorescent in the presence of this radiation.

One or the other of these methods is used for the recognition of pathologic conditions existing in the human tissues.

The fluorescent screen appears at first sight to be an easy way of recognizing abnormalities. Its value in the examination of the thorax, where the movements of the heart, lungs, and diaphragm have to be observed, is undoubtedly very great; but as an accurate means of recognizing any abnormality, it is untrustworthy. For instance, it is possible to fail to recognize simple transverse fracture of the tibia by its means. Its use is therefore to be deprecated in cases where great accuracy is necessary, and it is safer and better to make use of the more certain method, the photographic plate.

A further objection to the use of the screen is that the constant exposure of the hands and other parts of the body of the observer may result in an intractable, dangerous and chronic dermatitis.

By using a photographic plate the danger of dermatitis can be avoided, since it is not necessary to expose the hands at all; and at the same time greater accuracy is ensured and a permanent record is obtained.

Although examination by radiography is a somewhat tedious procedure in comparison with direct observation by the fluorescent screen, yet it is less difficult if the photographic side of this method is approached in a proper and businesslike manner.

Interpretation of Radiograms. A successful result in X-ray examination involves a clear understanding of the meaning of the radiogram produced. Even with the most accurate knowledge of anatomy, it is difficult to interpret X-ray shadows; for a radiogram is only a shadow, and the outline of the part thus demonstrated is liable to great variation. For example, in the case of injury to bone, it is always possible to secure strong and accurate X-ray shadows of the part, and no error ought to be made in diagnosis, yet errors of this kind are not uncommon.

To avoid such mistakes, it is imperative that the quality of the radiogram secured should be the best possible. For instance, in the examination of the ankle-joint and the bones of the foot, a radiogram which is flat, indistinct, and altogether wanting in detail, is of no value, while a radiogram of good quality of the same ankle-joint and foot, is of value. The interpretation of the latter is easy, while that of the former would be almost impossible, and certainly inaccurate.

The usual practice in securing radiograms is to place the subject in a position considered likely to give the best results, and then roughly, almost at random, to place the tube in some unknown relation to the part of the body under examination. The resulting shadow is often of no value because it is wanting in detail and depth. One method of avoiding this fault is to produce stereoscopic views of the part examined.

Two views having been secured in stereoscopic register, and placed in a stereoscope, the part can be viewed in relief. Theoretically, then, by this means one is able to view the parts of the body opaque to the X-rays as they would appear to the naked eye. In practice, however, this method, though it may prove of value in exceptional circumstances, is laborious. Moreover, though the parts may be made to appear in relief, they are not really as one would see them with the naked eye, but are still X-ray shadows.

A more practical method is to ensure that in all cases radiograms of any part of the body be absolutely comparable with one another by taking care to maintain the same relationship between the X-ray tube and the part under examination. For example, in making an examination of the ankle-joint, the limb is placed in a prescribed position, and the anode of the X-ray tube, that is, the actual source of the X-ray, is brought into accurate relationship to the tip of the internal malleolus by a simple mechanical contrivance, the details of which need not be dealt with here. This relationship between the tube and the ankle can always be reproduced, and therefore the shadow of a normal ankle-joint can always be obtained under the same conditions for comparison with the radiogram of the suspected ankle.

In this way, not only is the surgeon able to select the view of the part which will have the depth and detail necessary for proper interpretation, but, the shadow being familiar, he can more easily recognize any abnormality.

A radiogram secured under the conditions usually adopted, shows definite and known anatomic relationship between the bones and the X-ray tube, namely, with the anode of the tube directly opposite the tip of the internal malleolus.

To render this method of examination more perfect, there has been devised a system of radiography containing a definition of the relationships between the tube and the various parts of the body which have been found to give the most useful views, and also radiograms of the normal appearances of each part at the ages respectively of 5, 15, and 25 years.

By using this system the surgeon can secure a radiogram of any part of the body, of the requisite standard in quality, while he has at hand a normal radiogram of that part for comparison with the abnormal.

Having secured a radiogram of good quality, it is necessary for the purpose of interpretation that it should be viewed in a suitable light. The best for the purpose is a bright light shaded with opal in a dark room. The negative may be viewed at its best while still wet. Considerable loss of detail follows the taking of prints, which for this reason may greatly detract from the value of the radiogram.

It is a mistake to suppose that X-ray examination in the diagnosis of diseases can replace the older and well-tried clinical methods of investigation; it is merely a useful means of acquiring knowledge which, in conjunction with accurate clinical investigation, leads to a more accurate diagnosis and prognosis, and is often most useful by suggesting a more suitable line of treatment. It must be remembered that this method of investigation has been in use only a comparatively short time. In some diseases no definite statement is yet possible that may not prove in the future to be misleading.

At present the therapeutic use of the X-ray is rightly falling into the hands of the dermatologist and the medical clinician. In surgery, outside of the conditions mentioned above, its use is limited to lupus, keloid, epithelioma, sarcoma and carcinoma, both before and after operation.

CHAPTER XIX

DRESSINGS AND BANDAGING; SOLUTIONS AND OINTMENTS; SKIN GRAFTING

DRESSINGS

Dressings. These may be either dry or wet.

Dry dressings consist of gauze and bandage or of cotton and collodion (the cocoon dressing.)

The most convenient form in which sterile gauze can be obtained is in small squares in individual envelopes. Large packages are contaminated with the first opening and are inconvenient.

The cocoon dressing is occlusive and should never be applied over an infected area. It is applicable to sensitive areas for protection, and to operated areas not liable to infection.

Protective varnishes, such as collodion, compound tincture of benzoin, or pure ichthyol, are useful where little protection is indicated.

Wet dressings. Two distinct therapeutic actions may be derived from the wet compress, depending upon whether or not an impervious covering is employed. These actions are *antiphlogistic*and *hyperemic,* and these in turn may be either *antiseptic* or *astringent.* The wet dressing, without a covering, is cleansing and heat reducing, because of evaporation. There should be frequent replenishment of the solution in the treatment of any infected wound or where it is desirable to reduce inflammation.

A wet dressing with an impervious covering is contraindicated in the presence of pus, the warmth and moisture of such a dressing being congenial to the growth and to the multiplication of bacteria.

It is evident, therefore, that a wet dressing with an impervious covering can safely be employed only in conditions where the skin is unbroken, such as sprains and bruises.

The two general therapeutic actions, aside from those of causing hyperemia, are antiseptic and astringent. For the relief of pain and for the reduction of inflammation, wet dressings are the most effective form of treatment because (1) they are aseptic; (2) they permit free drainage; (3) no new granulations are disturbed in changing the dressing.

A great many different solutions are used and among these are:

1. sterile water;

2. ordinary saline solution (a teaspoonful of salt to a pint of water);

3. saturated solution of boric acid (prepared by dissolving a teaspoonful of boric acid powder in a pint of water);

4. Thiersch's solution (prepared by dissolving 15 grains of salicylic acid and 90 grains of boric acid in a pint of water);

5. Burow's solution (a solution of aluminium acetate prepared by dissolving 675 grains of alum and 270 grains of lead acetate in a pint of water.U.S.P. formula);

6. solution of bichloride of mercury (varying in strength from 1 to 3000, to 1 to 10000);

7. 2 per cent. solution of creolin or lysol;

8. U.S.P. lead and opium wash;

9. aqueous solution of ichthyol (varying from 5 to 50 per cent. according to the indications);

10. black wash (made by dissolving 64 grains of calomel in a pint of lime water—this solution only being used in luetic cases).

11. white wash (prepared by mixing zinc oxide, 2 drams, solution of subacetate of lead, 3 drams, glycerine, 4 ounces and lime water, 4 ounces);

12. Dakin's solution (hypochlorite of soda), prepared as follows:

chlorinated lime (bleaching powder)	200 gm.
sodium carbonate,dry	200 gm.
sodium bicarbonate	80 gm.

Put the chlorinated lime in a 12 litre flask with 5 litres of ordinary water and let stand over night. Dissolve the sodium carbonate and bicarbonate in 5 litres of cold water; then pour this into the flask and shake it vigorously for a minute and let it stand to permit the calcium carbonate to settle. After half an hour, siphon off the clear liquid and filter it to obtain a perfectly limpid product. The antiseptic solution is then ready for surgical use: it contains about 0.5 gm. per cent. of sodium hypochlorite with small amounts of neutral salts. It is practically isotonic with blood serum. Never heat the solution, and always keep it from the light. If in an emergency it is necessary to triturate the chlorinated lime in a mortar, do so only with water, never with the solution of the soda salts.

This solution has been used extensively abroad in the treatment of infections and wounds and has given splendid results.

(A proper quantity of Dakin's solution for office purposes would be about one-tenth of the prescription above given.)

DUSTING POWDERS

These are employed either as antiseptics or as astringents or for both purposes. Their use is limited, and they are employed only where the secretion is scanty.

Among the various powders used are: aristol, dermatol, boric acid, orthoform, calomel, protonuclein, zinc oxide, alum, scarlet red, etc.

Thymoliodide, or *aristol,* is a splendid antiseptic powder and enjoys the advantage over iodoform of being inodorous.

Iodoform should only be used in tubercular conditions.

Dermatol, or *bismuth subgallate,* combines the astringent and mildly antiseptic qualities of bismuth and gallic acid.

Boric acid is mildly antiseptic.

Calomel should only be used in syphilitic conditions.

Zinc oxide and *alum* are both astringent.

Scarlet red (5 per cent.) with boric acid (95 per cent.) is indicated for the stimulation of granulations.

Solutions. Among the various solutions used are silver nitrate, in various strengths, zinc and copper sulphate, ichthyol, balsam of Peru, nitric acid, sulphuric acid, trichlorand monochloracetic acid.

Silver nitrate is employed for its astringent action, as are also the *copper* and *zinc sulphates.*

Balsam of Peru is used for its stimulating action.

The stronger acids are employed for their escharotic qualities.

"Red wash" (made up from the following formula: zinc sulphate 20 grains, compound tincture of lavender 30 minims, distilled water to make 8 ozs.) has a powerful astringent action and promotes cicatrization, especially when there is a tendency for the granulations to become exuberant.

In the treatment of chilblains, a strong astringent is desirable to constrict the diluted capillaries.

The stronger *lotio alba* of the national formulary, containing equal parts of the saturated solutions of zinc sulphate and potassium sulphuret, is markedly astringent and has a drying effect upon the skin.

STYPTICS

Styptics. These may act either by causing clot formation in the cut arteries, or by causing the retraction of their edges. In the latter class are included such drugs as *hydrastine* and*adrenaline.*

The disadvantage of using these drugs lies in the fact that secondary hemorrhage is possible when their constrictor action is over. The styptics causing clot formation are therefore to be recommended.

They should be non-irritating, antiseptic, and styptic, at the same time. Such a preparation is practically unknown.

Peroxide of hydrogen on a pledget of cotton, placed over the bleeding area, may effect a clot formation.

The U.S.P. *liquor ferri subsulphatis*, better known as Monsel's solution, is the best and most effective styptic that we have. Monsel's solution, however, is not antiseptic and entrance of bacteria into the wound is possible, unless, it is applied with a sterile applicator or is dropped directly upon the wound from the bottle.

The U.S.P. *tincture of iodine* in equal parts of water, applied to the bleeding area may, besides sterilizing it, stop bleeding.

Should none of the above effect a stoppage of the bleeding, other means must be sought. A bit of sterile gauze pressed quite firmly against the area, should next be tried. If this fails, a wooden applicator, prepared with Monsel's solution may be employed. A cotton wound applicator, unless dipped into a strongly antiseptic solution, contains millions of bacteria from the fingers. The use of the ancient styptic stick of alum, copper or silver is discountenanced everywhere as uncleanly.

SOLVENTS

Solvents. Under this heading, those substances which are known to soften tissue will be considered.

Sodium hydroxide, up to a saturated strength, or an ointment of *salicylic acid*, 5 per cent. to 50 per cent., depending upon the density of the tissue to which it is applied, are the ones commonly used.

These two drugs have the power to macerate dry, hard tissues.

Experience is necessary for the proper use of tissue solvents as the length of time that they are allowed to act is of as much importance as the strength of the solution.

Sodium hydroxide solution can be instantly neutralized with any acid and for this reason is preferable.

OINTMENTS

Ointments. In the list of ointments, the much vaunted virtues of advertised compounds are usually found.

Ointments and oils are used in the treatment of wounds and ulcers, either to stimulate granulations or to soften thick epidermis.

Ointments should never be used where there is a profuse discharge, as eczema is a complication which very often follows such treatment.

A great many different kinds of ointments are used and among these are:

Sulphur in 10 per cent. strength, or *ammoniated mercury* up to 5 per cent., where a paraciticide is indicated.

Balsam of Peru in 10 per cent. strength for the stimulation of granulations; or *balsam of Peru* and *castor oil*, equal parts; also *boric acid*, or *ichthyol* for their antiseptic properties.

Ten per cent. *mercurial*, for syphilitic cases.

Lassar's paste (which consists of salicylic acid, one dram, starch and zinc oxide, each one ounce, and vaselin to make 4 ounces) is used when there is an eczema present.

One of the oldest as well as one of the best applications is balsam of Peru, which has a powerful effect in increasing the growth of granulations, but often after this has occurred the granulations are apt to become exuberant with little tendency to cicatrization.

The ointment which has given the best results is *scarlet red*, an aniline dye, which is known chemically as a sodium salt of a disulphonic acid derivative. Scarlet red (Biebrich) was originally prepared as a dye for wool and silk, and is so named because of the fact that it was first manufactured in the town of Biebrich. It was first used for medicinal purposes in 1907 in an 8 per cent. strength; because this strength was found to be too irritating, it was alternated with a bland ointment every 24 hours. It is now used only in strengths varying from one-half to five per cent., for the latter has proved to be as strong as necessary. When applied to granulating surfaces, scarlet red is sometimes absorbed in sufficient amount to color the urine a bright red, and a number of acute cases of nephritis have been reported from its use.

Its application to granulating surfaces causes healing, not by the formation of scar tissue, but in every case by producing a high grade of normal skin (this can be demonstrated by sections), which very soon becomes freely movable on the underlying tissue. The return of sensation in the healed area takes place from the periphery inward, instead of upward from the underlying tissue.

Scarlet red ointment should be applied in the following manner: after thorough cleansing of the part with tincture of green soap and water, then ether and finally 93 per cent. alcohol, the ointment should be spread in a thin layer over the entire surface on a piece of sterile gauze, and over this an ordinary dry sterile dressing. If the ointment is applied too thickly it may cause granulation tissue to break down, and for this reason it should be spread in a thin layer upon the granulating surface or its edges. Usually the dressing should be left undisturbed for from 24 to 48 hours, then reapplied, as indications warrant. The patient should invariably be informed that the dressing will be stained red, so as to forestall unnecessary alarm, due to the belief that a hemorrhage has occurred. He should also be apprised of the fact that stains on the linen are hard to eradicate. In removing the dressing, if it is adherent to the granulations, some peroxide of hydrogen should be used to loosen it. The skin about the granulating surface is best cleansed by benzine as this removes all traces of scarlet red better than any other solution. The three formulas that are recommended are the following:

	Strength	
	Grains.	Percent.
Scarlet red (medicinal Biebrich)	15	1
ungt. acidi borici q.s., ad. 3 ounces.		
Scarlet red (medicinal Biebrich)	45	3
ungt.zinci oxidi q.s., ad. 3 ounces.		
Scarlet red (medicinal Biebrich)	75	5
balsam Peru, 75 minims.		

Petrolati q.s., ad. 3 ounces.

The first is indicated where its use is desired over a large area and for a long time; the second, where an astringent action is required because the granulations are profuse; the third, where the granulations are sluggish and require stimulation.

The ointment in a 10 per cent. strength is not recommended because it is too irritating.

In cases of chronic leg ulcers, especially those associated with enlarged veins, it is impossible to effect a cure until the chronic congestion of the limb is relieved and the blood supply of the part approaches the normal.

Often all that is necessary is a gauze, muslin or flannel bandage, properly applied over the dressing and extending from the ankle to the knee.

A rubber bandage when applied with moderate, even pressure, has for its purpose the relief of congestion, but in a great many cases the rubber has an irritating effect on the skin.

When the granulations are almost on a level with the surrounding skin, and also when there is considerable thickening of the edges of the ulcer, the best means of keeping up an even pressure and causing absorption of the thickened margins, as well as of hastening the epithelial growth, is to apply zinc oxide adhesive plaster in strips, one-half to one inch in width. These strips should overlap to the extent of about one-third of their width; should extend about three-fourths of the way around the limb, and should be evenly and smoothly applied. They should be started about one inch below the ulcer and should run from two to three inches above it.

BANDAGING

Bandaging of Leg. The final stage after the dressing has been put on, consists in the application of the bandage. A bandage possesses advantages over strapping in being less irritating to the skin; in being more quickly put on and taken off; in being more easily removed without disturbing the surface, and in more completely allowing the formation of the granulations.

The bandage is also superior to a laced stocking, as the latter does not properly embrace the foot.

The bandage material can be either gauze, muslin or flannel. The last is considered the best because this material is thin, yielding and elastic and yet almost any degree of compression can be exercised with it.

In edematous swelling in general, the flannel appears very suitable, as it is soft to the skin and accommodates itself to the greater or less distension of the limb, arising from the increase or diminution of the fluid. The bandage should be at least six yards long, if required for an ordinary adult, and the width should be from two to three inches. Every portion of the limb, from the toes to the knees, should be equally and evenly compressed. Compression is of such absolute importance that without it everything else will be comparatively ineffectual. This being so, very much will depend on the manner in which the bandage is employed.

Without practice, it is not easy to properly apply a bandage to the leg, and probably this difficulty is the chief reason why preference is often given to adhesive plaster, as this sticks wherever it is put.

The blistering and excoriation often produced by strapping, and the time consumed in its application, are sufficient reasons for acquiring skill in the art of bandaging; an art whose comforts and advantages are appreciated by the patient.

Before using, the bandage should be rolled up very tightly, so that it may be grasped easily and held in the hand firmly without slipping. In putting it on, unwind only that portion which is being applied to the limb, because if it be loose in the hand, or if a considerable piece be unrolled at a time, it cannot be applied firmly or smoothly. The bandage should always be carried up to the knee, even if the ulcer or wound be seated on the lower part of the leg or on the foot itself, as the object of its application is not merely to cover the ulcer but also to support the vessels of the limb. If the bandage be discontinued on any part of the leg, it is liable to become loose and fall down.

It is desirable also that the patient should not wear a garter above the bandage, as anything unequally tight in the course of the veins is calculated to obstruct the free passage of the blood.

The firmness with which the bandage is put on is, of course, chiefly for the purpose of gaining the good effects of compression on the structures beneath, but besides, it contributes very much in making the bandage remain in its position when applied. Encircle the limb with it in a loose, careless manner, and it will fall down almost immediately the patient begins to walk about. Tight bandaging is extremely well borne if performed in a complete and methodical way, beginning at the lowest portion of the foot around the first joints of the toes and ending just below the knee.

The proper application of the bandage is of such great importance, especially in the treatment of varicose ulcers of the leg, that it should, when possible, always be done by the doctor himself. It is difficult for the most skilled layman to put a bandage on his own leg. The real practical difficulty lies with those patients who live at a distance from the doctor and who can only visit him once a week or at ten day intervals. These must be taught to dress and bandage the limb, and generally some friend or relative will learn to superintend the details.

The length of time which elapses before the bandage and dressings are removed and reapplied must necessarily be determined by the circumstances of each case. When the ulcer is very extensive and the discharge proportionately great, it may be advisable to dress the leg every day at the beginning of the treatment. Generally speaking, an ulcer of the leg is disturbed too often. To take off a dressing and put on another, even though done with the greatest care, interrupts the healing process and the natural steps to cure. Let the dressing remain on until some uneasiness points to the propriety of taking it off, for the purpose of allowing the escape of the discharge. Delay the removal of the dressings as long as possible without carrying the forbearance too far. Avoid extremes of waiting too long or of meddling too soon. Taking the average case, an interval of three days may in general be safely permitted.

Spiral Bandage of the Great Toe. In applying this bandage, the initial extremity of the roller is secured by two or three turns around the ankle and the bandage is carried obliquely across the dorsum of the foot to the base of the toe to be covered, and next to its tip, by oblique turns; a circular turn is then made and the toe is covered by ascending spiral or spiral reverse turns until its base is reached, from which point the bandage is carried obliquely across the dorsum of the foot and finished by one or two circular turns around the ankle. The end of the bandage may be secured by a pin or may be split into two tails and secured by tying.

Spica Bandage of Great Toe. This bandage is applied by placing the initial extremity of the roller upon the ankle and fixing it by two circular turns; the roller is then carried obliquely over the dorsal surface of the foot to the distal extremity of the great toe; a circular turn is next made and the bandage is carried upward over the back of the great toe to the ankle, around which a circular turn should be made; ascending figure of eight turns are then made around the great toe and the ankle, each turn overlapping the previous one, two-thirds, and each figure of eight turn alternating with a circular turn around the ankle. These turns are repeated until the great toe is completely covered with spica turns and the bandage is completed by circular turns around the ankle.

French Bandage of the Foot. In applying this bandage the initial extremity of the roller should be fixed on the leg just above the ankle and secured by two circular turns around the leg; the bandage should be carried obliquely across the dorsum of the foot, to the metatarsophalangeal articulation, at which point a circular turn should be made around the foot; the roller should then be carried up to the foot, covering it with two or three spiral reverse turns; after this a figure of eight turn should be made around the ankle and instep; this should be repeated once to cover the foot, with the exception of the heel, and the bandage continued up the leg with spiral reverse turns.

Spica Bandage of the Foot. In applying this bandage, the initial extremity of the roller should be fixed just above the ankle and secured by two circular turns; the bandage should then be carried obliquely over the dorsum of the foot to the metatarsophalangeal articulation; a circular turn around the foot should be made at this point and the bandage continued upward over the metatarsus by making two or three spiral reverse turns; it should then be carried parallel with the inner or the outer margin of the sole of the foot, according as it is applied to the right or left foot, directly across the posterior surface of the heel, and from this point it should be conducted around the outer border of the toe and over the dorsum, crossing the original turn in the median line of the foot, thus completing the first spica turn. These spica turns should be repeated, gradually ascending, by allowing each turn to cover three-fourths of the preceding one, until the foot is covered, with the exception of the posterior portion of the sole of the heel; the turns should cross one another in the medium line of the foot and should be kept parallel throughout their course.

Bandages for the Foot and Leg. Whenever possible the patient should be kept in bed, or, at least, in the recumbent position with the leg elevated, but when circumstances do not permit of this the veins can be supported in various ways. Elastic stockings are excellent but expensive, and not durable. Bandages of rubber cloth, or woven bandages rendered elastic by the character of the mesh, or Martin's plain rubber bandage may be employed. The last named is put on smoothly but not too tightly, for in walking the leg swells, so that a uniform pressure is established. As the rubber prevents evaporation it acts like a wet compress, stimulating the granulations, but very often producing eczema around the ulcer. The rubber bandage should be washed carefully at night with soap and cold water and must be kept clean. In one patient a firm elastic stocking of vulcanized rubber will give the greatest ease and comfort, while in another the resulting irritation will prove unbearable. As regards the flannel bandage it has already been described at some length.

The essential feature of ambulatory treatment is a good dressing to prevent congestion, and Unna's paste is ideal for this purpose. The paste necessary for the bandage is prepared as follows: first dissolve four parts of the best gelatin in ten parts of water by means of a hot water bath. While the fluid is hot add ten parts of glycerine and four parts of powdered white oxide of zinc; stir briskly until the mixture is cold. Another formula for the paste, and the one recommended, consists of the following: white gelatin, 2-1/2 ounces; water, 8 ounces; zinc oxide, 2-1/2 ounces, and glycerine, 4 ounces; prepared as above. The paste should always be melted before use by placing the receptacle in a hot water bath or in an ordinary copper sterilizer, such as that employed for boiling instruments. A small tin can be used, and a piece of paste about four inches square is cut into fine pieces and put in the can. This is placed in the sterilizer, into which is poured water to a depth of about two inches, so that the can is but slightly immersed. No top should be placed on the can. An ordinary stove or gas range can be used for heating purposes. A very important fact to remember is that no water is to be put into the can with the paste.

The leg is next cleansed, and after the paste has been thoroughly melted it is applied from the base of the toes to the knee, as hot as the patient can comfortably tolerate it, by means of an ordinary small paint-brush. Then a layer of gauze bandage (two to three inches in width, according to the limb) is applied, then a layer of paste, and in this manner two or three thicknesses of bandage are used, depending on the case. In thin people, it is necessary to use only one or two layers of bandage, whereas in stout persons several layers may be required. After the last application of the paste, some non-absorbent cotton is spread on the bandage, giving it the so-called "moleskin" plaster finish. Another way of finishing the dressing is to dust some ordinary talcum powder on the last layer of the paste, giving the bandage the appearance of a plaster-of-Paris dressing. If there is an ulcer, a window can be cut out, thus providing for the drainage of the secretions. The length of time this dressing should be left on depends on a number of conditions, especially the amount of secretion, and whether the patient has to remain on his feet very much. Ordinarily, the bandage can remain on for one week, but indications may be such that it need not be removed sooner than the tenth day, and in some instances it can be kept on for three or four weeks. To remove it, an ordinary bandage-scissors is used to cut the dressing, and it peels off without disturbing any of the granulations on the ulcer.

PROMOTION OF NEW EPITHELIAL GROWTH AND CICATRIZATION

The value of nitrate of silver and red wash as stimulants of the healing process has already been mentioned. They are also of value in producing cicatrization and in promoting the covering of new epithelium over the ulcer or wound. If the solid stick of nitrate of silver be applied very lightly to the edges just inside the pale bluish line of advancing epithelium, so as to produce a white film on the surface, this slight cauterization will be found to aid in strengthening and cornifying the new, delicate and previously invisible epithelial cells and in preventing them from being washed away by the discharge from the ulcer. The solid stick of nitrate of silver is also of benefit in destroying the exuberant granulations which project above the surface of the surrounding skin; often, by piercing these flabby granulations in several places with the solid stick held perpendicular to the surface, cicatrization is hastened. After the granulations are level with the surrounding skin the covering of the ulcer or wound with new epithelium is hastened by the application of some smooth surface along which the epithelium can spread. For this purpose zinc oxide plaster or some thin rubber may be used.

In some old chronic cases, healing is prevented by the fact that the base of the ulcer cannot contract owing to its being bound down by fibrous scar tissue. This binding down of the base and edges of the ulcer also tends to cut off the blood supply, and therefore in this additional manner healing is hindered. For the relief of this condition a number of procedures have been devised. Mattress sutures, introduced through the normal skin beyond the edges of the ulcer and passing beneath it, out through the skin on the other side, is one method. By tightening these sutures, over a button or metal plate, the ulcer can be lifted from the underlying tissues. Another method, called "starring of the ulcer," consists in a series of radiating incisions through the base and edges of the ulcer, the part from which the incisions radiate corresponding with its centre. In this and in the following operations, in order to obtain a favorable result, it is necessary that the incisions pass completely through the cicatrical tissue which forms the base and edges of the ulcer into normal tissue. "Cross-hatching" of the base of the ulcer by means of a series of incisions at right angles to one another, and at a distance of about one-half inch apart, is often of value in aiding the healing of a chronic ulcer, the continued existence of which and failure to heal having been due to its thickened, adherent base and edges. Circumcision of a chronic ulcer consists in making a circular incision around it through the normal skin. A modification of this method consists in making a series of overlapping, short, curved incisions surrounding the ulcer, instead of a single circular incision. In these last two methods it is necessary that the incisions be made through normal skin, and that the wounds be made to gape, if necessary, by packing them with gauze.

When the ulcer or wound is of considerable size, it is often impossible to secure healing even by these methods. It may for a time appear as if it were going to heal, and a pale blue line of newly formed epithelium may spread out from the edges, but instead of the epithelium continuing its progress, at a subsequent dressing it will be found to have disappeared. In these cases, as well as in those in which the size of the ulcer would necessitate a long delay for a cure or in which the subsequent contraction of the scar would produce deformity, skin grafting, skin transplantation, or some form of flap operation is indicated.

SKIN GRAFTING TO OBTAIN A SOUND SCAR

A very important object in the treatment of all ulcers is to obtain a sound scar. In ulcers affecting the lower extremity in elderly people, the scar resulting from spontaneous healing is weak and readily breaks down if the patient does much standing or walking. The patient is therefore frequently obliged to give up work in order to get the ulcer re-healed, or must be content to employ means which merely prevent its extension and relieve some of the discomfort. When the best possible scar is desired, and when it is important to avoid marked contraction, it is necessary to adopt some method of skin-grafting.

There are three plans by which rapid healing of an ulcer may be brought about: Reverdin's epidermis grafting; Thiersch's skin grafting, and the use of the whole thickness of the skin.

Reverdin's Method. In this procedure small thin portions of the superficial layer of the skin are snipped off with a curved scissors. Pieces about the size of a hemp seed are planted on the surface of the granulations at short distances from one another. Epidermic growth occurs from each of these little points, and the result is that numerous small islands of epithelium form over the surface of the ulcer. If the grafts be close enough together and the conditions be favorable to healing, these islands soon coalesce and thus rapid cicatrization is obtained. The grafts should not be too far apart, because they appear to have only a limited power of reproduction.

With a view to obtaining a sounder scar, thicker and more extensive portions of the skin must be taken and the grafts must be applied close together. There are two ways of doing this: either by using the whole thickness of the skin or by employing Thiersch's method, in which about half the thickness of the skin is shaved off.

The procedure where the whole thickness of the skin is employed need not be described, partly because the results are not satisfactory and partly because all the conditions for which it was introduced are better fulfilled by Thiersch's method.

Skin grafts may be taken either from the patient himself or from another individual. When the patient is much debilitated, the cutaneous epithelium shares in the general malnutrition and under these circumstances a graft from a healthy subject might succeed better than one taken from the patient.

Thiersch's Method. In employing this method the skin which is to be used for the grafting must first be shaved and disinfected in the usual manner, as has been previously described. The presence of hairs on the grafts seems to interfere materially with their union.

Preparation of the Ulcer. *Preliminary.* It is of no use to graft a sore which is actually ulcerating; it must be brought into a healthy condition, and healing must have commenced before transplantation is likely to be successful. The best criterion that healing is taking place is the presence, at the edges, of the dry line which indicates recently formed epithelium. Some surgeons wait for a considerably longer time before grafting in order to get a firm layer of granulations, but experience shows that it may be safely resorted to as soon as healing begins around the edge. A second essential is that the ulcer shall be clean. If the discharges be septic, the graft, which is, after all, merely a piece of dying tissue, will become impregnated with decomposing pus and may rapidly become loosened, die, and undergo decomposition. The methods of rendering the ulcer aseptic have already been described.

Operative. The following is the method of procedure: after the patient has been placed under an anesthetic, the granulations over the whole surface of the ulcer are forcibly scrubbed off with a firm nail-brush, or are evenly scraped away, taking care, however, to remove only the soft layer of granulations and not to go through the deeper one of newly formed fibrous tissue into the fat. A surface is thus left which is smooth, highly vascular, and firm, and which consists of the deeper layers of granulation tissue that have already become organized into fibrous tissue. In cases of ulcer of the leg it is also advisable to remove those portions of the edge which have already become covered with new epithelium. If the transplantation be limited to the parts actually unhealed, the result is disappointing as a rule, for while the part grafted remains sound, the margin where spontaneous healing had occurred, is apt to break down, and thus a narrow line of ulceration appears at the edge of the ulcer.

After the layer of granulations has been removed and the newly healed edge of the ulcer has been cut away, the bleeding must be arrested completely before the grafts are applied. The most rapid method is to pour a few drops of adrenalin chloride (1 to 1000) solution over the raw surface, when the oozing ceases immediately. If adrenalin be not at hand the following plan will be found satisfactory: any spouting vessel is clamped and a large piece of sterilized gauze or thin sheet rubber is applied over the raw surface of the wound; outside this, several sponges are placed and a sterilized bandage is bound firmly over them. If the sore be small and an assistant be available, he may apply the pressure. Pressure is employed indirectly through the protective in this way, because if it were made directly upon the surface of the wound by means of the sponges, bleeding would recommence when the latter were removed, as they stick to the raw surface.

While the bleeding is being arrested the surgeon cuts his skin grafts from any part of the body, as he thinks fit As a rule they are taken from the front of the thigh, but the side of the abdomen may be selected. The area from which the grafts are to be cut is disinfected, and the surgeon grasps the limb from behind with his left hand in such a way as to make the skin over the front of the thigh as tense as possible; in doing this he pushes the soft parts well forward so as to make the anterior aspect of the limb as flat as possible. The skin is further put on the stretch vertically by an assistant, who pulls it upward and downward. These precautions are important, as without them it is almost impossible to cut a graft of even width. The razor, which should have a very broad blade, is dipped into a boric acid solution and is kept constantly wet with it whilst the grafts are being cut. Unless this be done, the graft adheres to the blade and may be either partially or wholly cut through before a sufficient length can be obtained. The razor is made to penetrate through about half the thickness of the skin, and then, by a lateral sawing motion, the grafts are cut as broad and as long as possible. After a little practice it is easy to cut them about two inches in breadth and about four or five inches in length.

If one graft be insufficient, it is best to slide it off the razor and leave it on the bleeding surface; in this way it is kept warm and moist. Some surgeons put the graft into warm saline solution, and it is said to then spread out more easily afterwards. Small skin grafts can be cut under local anesthesia.

Application of Grafts. When a sufficient number of grafts have been cut, the bandage, sponges and protective are removed from the raw surface of the ulcer and the grafts are applied to it if the bleeding has stopped, as is generally the case. The raw surface usually has a thin layer of blood-clot upon it, and this should be wiped away.

Each graft is lifted with forceps or the fingers and applied with the cut surface downward, and then is carefully unfolded by means of two probes and stretched evenly over the surface. The grafts should overlap the edges of the skin and also each other, so that no part of the raw surface is left exposed, for granulations always spring up on the uncovered parts and are apt to destroy the grafts in their vicinity; moreover, a thin scar is left at these points which may break down subsequently. The graft is always thinner at its edges than at its centre, and it is these thin edges which overlap each other or the margin of the skin; there is no real sloughing of these overlapping portions.

The dressing should be left on the grafted surface for about five days; in some cases even for a week. If the wound be aseptic, no suppuration or decomposition takes place beneath it. Before being removed, the dressing should be thoroughly soaked with a 1 in 2500 sublimate solution, for otherwise it may stick at the edge and adhere to the graft, which may thus be peeled off, unless great care is taken. The parts should be gently cleansed with the same solution, and a dressing similar to that put on originally should be employed for about another week. At the end of that time the grafts are fairly, firmly adherent and then a 5 per cent, boric acid ointment is the best application.

It will be found that even at the first dressings the grafts present a pink color and are adherent to the deeper surface, though they are still readily detachable. In the course of about a week the old cuticle peels off, but no raw surface is left. Later on, there is a great tendency to the formation of new epithelium, cornification, and drying-up, and it is to avoid the latter condition that ointments are so useful; in fact, until the scar is absolutely sound, it is well to keep the surface covered with some greasy application, the best being the 5 per cent, boric acid ointment.

For many months the grafted surface is likely to scale or crack, and this might prove a starting-point for the occurrence of sepsis which would cause the newly grafted area to slough. It is important to keep the scar as supple as possible, and therefore it should be constantly anointed with cold cream, vaselin, or lanolin. Grafted surfaces upon the face, however, do not manifest this tendency for any length of time.

Time Required for Cure. It is important to know when the patient may be allowed to walk about after an ulcer of the leg has been skin-grafted. If he begins too soon, the grafts will almost certainly become detached. That this will be so is evident from a consideration of the mode by which the adhesion of the grafts takes place. At first they adhere to the surface of the sore, simply by means of the effused and coagulated length. Cells rapidly spread into this length and in the course of two or three days the space between the grafts and the raw surfaces is occupied by a mass of young cells. In this tissue, new blood vessels develop and penetrate into the graft, whilst, at the same time, the cells of the latter grow and assist in the development of the young tissue and of the blood vessels. Thus the graft becomes vascularized; but for a considerable time the tissue between it and the surface of the sore contains many young blood vessels with delicate walls, and therefore, if the patient stands erect and allows the pressure of the column of blood to fall on these vessels, they rupture, and bleeding occurs beneath the graft and leads to its detachment.

It requires a long time before the graft is firmly incorporated with the tissue beneath by the development of elastic fibres; indeed, it may be reckoned that this union is not complete until from three to six months have elapsed. The graft will, in all probability, be destroyed if the patient walks about within three months of the transplantation. Hence, unless that time can be devoted to the treatment, it is not worth employing skin-grafting for ulcer of the lower limbs. By this, however, it is not implied that it is necessary to keep the patient in bed for the entire time, but merely that the foot must not be allowed to hang down, nor must any weight be borne upon it.

At the end of about six weeks the patient may be allowed to get up and lie on a sofa or sit with the leg on another chair, but the limb must not be permitted to hang down. After about three months he may be allowed to get about, but in order to prevent the detachment of the grafts, he should be fitted with a knee-rest and peg on which he walks, the leg projecting out behind him. If possible he should not put his foot to the ground until six months have elapsed. In cases of sores on other parts of the body, when the erect posture does not cause congestion of the part, the patient may be allowed to walk about after the first three weeks.

Results. The scar which results after skin-grafting performed in this manner is of a satisfactory character, and ulcers which have been intractable for years may be closed satisfactorily by this means. In order to obtain anything in the nature of a permanent cure, however, the prescribed period of rest must be adhered to rigidly.

CHAPTER XX

LOCAL ANESTHESIA

History. From Corning we learn that the ancient Assyrians alleviated and even entirely prevented the pain incident to circumcision by compressing the veins in the neck. Unconsciousness was probably induced in this way together with pressure on the carotids.

In India, centuries ago, the effects of opium and of Indian hemp were known and employed, and the ancient Egyptians were also conversant with the soporific effects of many drugs. We learn, from the same authority, much which he gathered from literature about the history of local anesthesia, and it is from Corning's well-known book on local anesthesia that most of this history is quoted.

In Peru, the Spanish conquerors learned that the coca loaf was held in high esteem by the natives, inasmuch as they observed that it was chewed by the high priests and nobility only, the vulgar being denied this privilege except as a reward of great merit or of distinguished valor. The leaf was regarded with awe and superstition and was supposed to possess supernatural powers. After the fall of the Incas, the Spanish not only permitted but encouraged the general use of the leaf in order to obtain more work from the natives, a result which the drug seemed to effect. It was also a source of great revenue to them and was sold at exorbitant profit to the natives who became enslaved to its effects but were able to endure great hardship while under its influence.

Chemists throughout the world, recognizing the potent action of the coca leaf, were soon engaged in the effort of extracting its active principle.

In 1859, after many had tried and failed, cocaine was evolved from crude extractives. Authorities differ as to whether it was Mann or Neimann, a pupil of Woehler, who first presented cocaine to the chemical world; however, fifteen added years elapsed before practical use for it was found. In 1862, Professor Schraff discovered that the tip of the tongue was rendered numb, and insensible when a little of the cocaine alkaloid was applied to it and that it remained so for a considerable length of time. Significant though this experiment was, the action of cocaine on the nerve-filaments was not recognized and the matter was not followed up until Dr. Karl Koller, of Vienna, began his experiments which resulted in a universal awakening to the use of a substance which, though known, had been allowed to remain unnoticed for ages.

Its anesthetic effect upon the eye was demonstrated by Koller at the Opthalmologic Congress at Heidelberg in 1884. Dr. H. D. Noyes was first to direct the attention of the American practitioners to Koller's results in the use of the drug. Its introduction was one of the greatest triumphs of modern surgery. It makes possible the discard of the systemic anesthetics in all minor surgical operations and also in many operations of considerable magnitude.

In the laboratory of Professor Stricker, Koller experimented on the eyes of a number of animals and thus reports his findings:

"A few drops of a watery solution of muriate of cocaine dropped on the cornea of a guinea pig, rabbit, or dog, or instilled into the conjunctival sac in the ordinary way, caused, for a short time, a winking of the eyelids, evidently in consequence of a slight irritation. After one-half to one minute the animal again opens its eyes which gradually assume a staring look. If now the cornea is touched with a pin head (in which experiment we have carefully avoided touching the eyelashes), the lids are not closed by reflex and the eyeball does not move, the head is not thrown back as usual, the animal remains perfectly quiet, and, on application of a stronger irritation we can convince ourselves of the complete anesthesia of the cornea. In this way I have scratched and transfixed the cornea of the animals used for experiment with needles, and have excited them with electric currents so strong as to cause pain in my fingers, and to become quite intolerable to the tongue. I have cauterized the cornea with the nitrate of silver stick until it became milky white; during all of this the animal did not move. The last experiment convinced me that the anesthesia involved the whole thickness of the cornea and did not affect the surface only. But if I incised the cornea, the animals manifested intense pain, when the aqueous humor escaped and the iris prolapsed. I have been unable hitherto to decide, by experiments on animals, whether or not the iris could be anesthetized by dropping the solution into the corneal wound, or by prolonged instillations into the conjunctival sac; for experiments to test the sensibility of non-narcotized animals are very complicated and difficult and do not yield unambiguous results. The last question which I subjected to experimentation on animals, viz., whether or not the inflamed cornea could be anesthetized by cocaine, was answered in the affirmative. The cornea in which I had incited a foreign-body-keratitis, became as insensible as a healthy one.

"Complete anesthesia of the cornea from the use of a two per cent. solution lasts ten minutes on an average. After such successful experiments on animals I did not hesitate to use cocaine also to the human eye, trying it first on myself and on some of my friends, and then on a great number of other persons, obtaining, without exception, the result of a perfect anesthesia of the cornea and conjunctiva."

Soon after Dr. Koller's report appeared, cocaine was used for a great many operations upon the eye, and its application to mucous membranes in general was soon taken up by practitioners everywhere.

Rectal, vaginal, otologic, rhinologic, oral and urethral anesthesia were soon found to be easy of accomplishment and many operations in these fields were performed under cocainization. The hypodermic injection of cocaine was experimented with and reported upon in 1884 by Drs. N. J. Hepburn, R. J. Hall, and Halsted.

PHYSIOLOGIC EFFECTS

Nerve Pressure; Anemia. That motor and sensory paralysis followed pressure upon a nerve has been well known for many years, and this has been utilized in the effort to produce anesthesia, artifically by applying a rubber tube or bandage around a finger or extremity, with the hope that "ligation anesthesia" would follow the arrest of circulation. This, however, has been unsuccessful as all that was thus accomplished was a slight sensation of numbness with no arrest of the sense of pain. This method could only be successfully carried out, were the nerves themselves subjected to sufficient pressure to injure them. Return to normal sensibility and motor function could not be expected for months.

Cold. The addition of common salt to ice hastens its liquefaction and consequently renders the mixture more cold. This knowledge has been applied in a method of producing anesthesia of limited areas of the skin. A gauze bag of the correct shape and size is filled with salt and ice mixed, and applied to the area to be anesthetized.

This method was used as far back as 1848, by Arnott, but was soon improved upon by Richet and others who used ether or rhigolene sprayed on the part to be anesthetized. It was found that extremely low temperatures could be obtained in this way, especially if a current of air were blown across the field of operation to hasten evaporation, and that a good local insensibility could be brought about if the circulation of warm blood could be either stopped or retarded with an Esmarch bandage or tourniquet. The method of obtaining local anesthesia through the agency of cold was found to be best accomplished by ethyl chloride and this substance is used in preference to any of the others previously mentioned, at the present time. Some years ago Dr. Martin W. Ware of New York experimented with both ethyl chloride and ethyl bromide and he found that the former was more serviceable in producing local anesthesia.

The Sensibility of Various Tissues. Karl G. Lennander, of Upsala, Sweden, shortly before his death, completed a chapter on local anesthesia for Keen's "Surgery" in which is set forth an elaborate account of the sensibility to heat, cold, pressure, and pain of the various nerve terminals throughout the body. In this great work he has given the world the results of many experiments on living tissues, experiments investigating the degree and kind of the tissues sensibilities; thus it is learned that "all internal organs receiving their nerve supply only from the sympathethic nerve and from the vagus, below the branching-off of the recurrent nerve, have no sensation, and that the abdominal and pelvic viscera are devoid of nerves to convey the sense of pain, heat, cold, or pressure."

From the same authority we are taught that the parietal peritoneum is highly sensitive but that the visceral covering is devoid of all sensibility, enabling the operator much freedom of manipulation within the abdominal cavity.

In a work of this limited size the sensibility of the various tissues cannot be fully treated but it should be borne in mind that the integument and the subcutaneous tissue, fat and muscles as well as the tendons, their sheaths, the muscles and periosteum and perichondrium covering the bones and cartilages throughout the body, are all highly sensitive to pain. It is also equally true that the bone substance, the bone marrow, and the cartilages are devoid of any of the four modalities of sensation. Articular surfaces covered with cartilage have no sensation, neither have the fibrocartilages any sensation.

GENERAL CONSIDERATIONS

Effect of General Anesthesia. Local or regional anesthesia is obviously the method of choice in all cases in which it is applicable. Not only is it desirable in the minor surgical operations and the more important ones upon patients suffering with a cardiac or nephritic derangement, where a general anesthetic is positively contraindicated, but in every instance where it is at all possible, the dangers and annoyances of general anesthesia should be avoided, and the regional or local anesthesia should be employed.

Among the advantages, aside from the number of assistants required and the discomfort immediately following the administration of a general anesthesia, are the absence of remote ill effects of the invasion throughout the entire system of a noxious chemical substance and its direct deleterious effects on many large organs such as the lungs, heart, kidneys, and liver, and the assurance, when a proper drug, dosage, and technic are employed, that death cannot be ascribed to the anesthetic.

Of remote ills of general anesthesia no estimate can be made, but that they are legion and of great severity is established. Deaths from general anesthetics in persons apparently able to bear them well, are extremely numerous. It has been estimated that one in fifteen thousand succumbs from ether anesthesia and this number would probably swell greatly were it possible to obtain the exact figures. Even this minimum of danger does not exist in local anesthesia.

An accurate knowledge of the neural anatomy of a particular region enables the operator to anesthetize large areas and to operate with entire freedom from the necessity of observing the appearance and conduct of his patients, many of whom, notably the alcoholic ones, behave badly, become cyanotic and breathe intermittently when under the effects of inhalation anesthetics. The absorption into the body of the substances employed by inhalation may also exert a baneful influence by reducing the powers of resistance upon an economy already lowered by disease, and also by retarding convalescence.

Advantages of Local Anesthesia. In minor or trivial affairs the elimination of pain is not to be considered lightly, for every patient, even the strongest, will appreciate anything which will expedite a cure and at the same time will relieve him of suffering. Rather than lose time from their work or suffer the nausea and dangers of general anesthesia, these patients often bear for years conditions which could easily be cured by operations under local anesthesia. In this class one must first think of hemorrhoids; of cysts; of fatty tumors; of foreign bodies in the hands and feet; of verruca and of ingrown nails. These conditions would be promptly relieved were the element of pain in surgical interference not to enter as a factor.

With a perfect technic, local anesthesia can also be employed with entire satisfaction for certain major operations, where the subject is suitable. Thus, herniotomies are performed with entire success, especially those cases complicated by strangulation in which the dangers arising from fecal vomiting and inspiration pneumonia, are greatly decreased by omitting the general anesthesia.

In many of the more severe conditions not to be classified as minor surgery, the surgeon may consider the comfort of the patient and his own convenience and employ local in preference to general anesthesia, even tho the patients may be of the most robust type.

In this group may be mentioned benign tumors at any visible part of the body, hernias, many scrotal and anal diseases and some conditions peculiar to the extremities, such as varicose veins. These conditions lend themselves kindly to local insensitization.

In certain emergencies where an operation must be performed immediately, such as tracheotomy, thoracentesis and strangulated hernia, local insensibility is imperative. In these operations local anesthesia is also more desirable because of the ill effects of vomiting, which are thus eliminated.

Weakness of the patient enters also as a demand for the exhibition of a local anesthesia in such operations as resection of a rib for empyema, in which instance the action of the heart or lungs is embarrassed. Other operations performed under local anesthesia for the same reason (weakness of the patient) are the exploratory operation for a probable inoperable cancer and the palliative operations such as gastrostomy, enterostomy and colostomy.

SOME VALID OBJECTIONS TO THE USE OF LOCAL ANESTHESIA

There are, however, valid objections to the general application of local anesthesia and the cases for its use should be selected with care. It does not produce relaxation nor does it give the surgeon perfect control over his patient. These are considerations which must be taken into account, especially in operating on patients of highly nervous temperaments. Though the patient may be convinced that he will suffer no pain, the mental attitude toward the local anesthesia, together with fear, may operate so strongly as to constitute a shock to the nervous system so great that a general anesthetic should be used and the local method abandoned, even were it apparently indicated.

Again, the injection of anesthetic drugs in cicatrical and inflamed tissues is quite difficult of accomplishment and because of the peculiarity of these tissues, diffusion throughout a given area is imperfect, hence insensibility is not complete.

The extravagant claims of enthusiastic advocates of this method of anesthesia have retarded its progress. Thus, in the hands of the competent operator it was given but a perfunctory trial to be discarded as impossible. At the present time, however, local anesthesia bids fair to become the method of choice, other things being equal, for many major operations not yet thus performed. Recent investigations alone these lines have developed methods of its application whereby it is possible to render insensible large areas of the integument, and regional anesthesia is performed by anesthetizing nerves proximal to the seat of operation, thus rendering amputations feasible.

A single element which has entered as a factor in retarding the progress of local anesthesia in general surgery, is that of regarding the operation as one fitted to the method rather than to the patient under consideration. It is obvious that this is a fallacy and the main issue in deciding between general and local anesthesia is: what will the patient best tolerate? In coming to a decision in the matter one should make a general survey and weigh first the general health of the patient; whether he be in perfect systemic condition or undermined by disease, whether the shock will be greater from one method than the other, and whether the part of the body to be operated on is one which will lend itself better to one method than to the other.

These elements are being and will continue to be considered as preliminary to operative procedure and in consequence, general anesthesia will cease to be given in a routine way.

GENERAL PRINCIPLES AND ESSENTIALS

The first essential to the successful production of local anesthesia is a proper equipment and one that is in good working order. Not only is it necessary to employ the best drug to this end but also to use a syringe having perfect mechanical construction and one not injured by boiling; as also needles of the length, lumen and shape suitable for the surface to be injected.

The old leather pocket syringes, on account of their not bearing water at high temperature without deterioration, should not be employed; this applies also to that variety of glass barreled metal-mounted syringe in which the glass is screwed into the metal end pieces.

The best syringes are those made of all metal or of all glass, the latter being preferred because one may see the contents and express out the air before injecting. Syringes of this type, because of the accurate fitting piston, must be thoroughly dried out after use, as the piston may stick fast within the barrel. All-glass or all-metal syringes must be selected with care as they are often imperfect, the calibre of the barrel being unequal in different parts of its length causing the piston to fit tightly in some parts, and thus to work with difficulty; and in other parts fitting loosely, allowing the fluid to escape backwards.

Syringes are also made in various sizes and shapes to meet certain requirements. For the edematization of large areas of loose tissue, where a considerable amount of a weak solution is intended, the use of a large barreled syringe will be found to save time and the annoyance of refilling.

For such work a five or ten c.c. syringe would be the most useful. The ordinary hypodermic syringe is about of two c.c. capacity (thirty drops), and serves the purposes of every-day work. It does very well for the amount of an anesthetic solution employed in opening an abscess or in the removal of a small cyst or lipoma or papilloma.

A barrel, large in diameter, requires more pressure on the piston in its operation unless the needle employed is also correspondingly large. For this reason, if the tissue in which the solution is to be injected is not loose or cellular, it will be found better to use a syringe in which the barrel is long and narrow. Such is the shape of the syringe intended for the injection of the gums, the peridental membrane, and also for the periosteum, cartilage or bony cellular structure. A long instrument is also required for use in the large cavities of the body such as the mouth, the vagina, or the rectum. In these localities, an extension fitting is often required to lengthen the instrument sufficiently to reach the desired part. It is also possible to attain this end by using a long needle; this, however, sacrifices rigidity.

For accomplishing the best results, the needles must also be selected for the work at hand. For the initial puncture in sensitive or inflamed tissue, it is proper to use a needle of the finest lumen so as to cause the least possible amount of pain. The ordinary needle, which comes with the usual hypodermic outfit, is about the proper length for the ordinary work already mentioned, but could be improved upon for anesthesia by being made a little finer in calibre.

This length (three-quarters of an inch) will be frequently found insufficient to reach the deeper tissues and in the removal of a more or less rounded growth, a longer needle must be selected at the start. Curved or angular ones are only needed in dentistry, where strength is also a consideration. Strength is afforded in those of short length by means of a reinforcement at the hub. Needles so augmented may also be of use in operations upon bone or dense structures in general; the curve, however, is not essential.

The surgeon should be fully conversant with the details of the operation which he is about to perform. His work should be definitely in his mind, for in operations under local anesthesia, there is no justification for a change of procedure after the beginning of the work. Account should be taken of the nature of the tissues to be anesthetized, for it is known that cicatricial tissues and inflammatory areas do not lend themselves to the action of these drugs. In a cicatrix, the diffusibility of the solution is impeded, and in an inflammatory or necrotic tissue, the changes in the quantity and quality of the fluids present, alter the action of the anesthetic.

In considering the personal element of the patient one meets a difficulty which is by no means minor, and full explanation for the selection of the local anesthetic with many assurances of the painlessness of the operation are frequently necessary. This is especially true with one of highly emotional temperament, and, to allay fear in such a patient is not always easy.

Whatever may be said regarding the mental state of the patient who is to receive an anesthetic, whether general or local, the surgeon must remember that to be calm does not always lie within the control of his subject, and it will be found that a hypodermic injection of morphine (gr. one-eighth to one-quarter) an hour before the start of the anesthetic, will often render possible the use of the injection method in a patient with whom it would otherwise have been impossible. Morphine injections, as suggested, are of advantage in patients on whom a major operation is contemplated; they loosen the musculature and diminish the sensations of parts not anesthetized.

The deliberate and confident manner and word of the surgeon go a long way in guiding the feelings of his patient, and a worried or apprehensive surgeon makes for a doubtful and sensitive patient, ready to cry out at the first prick of the needle. Therefore it is a part of good general technic for the surgeon to deport himself in a way conducive to cheerfulness, and conversation must be guided along these lines.

There are many who will writhe and groan at sensations (which they will admit later were not painful) incident to local anesthesia, such as the grating vibrations of instrumentation. Such a patient is not well fitted for the method and it is for the discerning surgeon to recognize such in advance, that he may operate under the most favorable circumstances.

Preparation of the Patient. Proper evacuation of the bowels and a stomach free of undigested parts of a previous meal, are desirable. The subject of an anesthetic should not be purged or starved as these are weakening processes and also disturb the tranquility so essential to a perfect anesthesia. The skin should be prepared so as to accomplish surgical cleanliness without irritating it so as to retard healing. It was once thought that soap, water, alcohol, ether and bichloride were absolutely necessary to this end. It has, however, been found that iodin, applied in the ten per cent. tincture to the site of incision, fulfills every requirement. Where shaving is necessary, it should be done first. In operations about the anus and scrotum, iodin is contraindicated because of its irritating properties; it is painful in these parts and dermatitis is frequently the result of its use.

Instruments. The instruments should be prepared and ready before the anesthetic is given, regardless of the form of anesthesia employed. The surgeon's hands should be rendered aseptic, no matter how trivial the procedure before him, and every precaution should be taken to guard against infection, which is always possible in any surgical procedure however insignificant.

Technic. Various methods of accomplishing the insensitization of a part may be employed. Thus, if the skin alone is to be incised, it alone will require injection and by careful insertion of the end of the needle it may be kept just under the epidermis, thus injecting the anesthetic endermatically in and about the papillae of the papillary layer.

Endermic Method. This method is an end-organ anesthesia, and the solutions employed are strong and act because of their drug content. It is not in any sense a pressure anesthesia. The skin should be picked up and pinched hard for the better insertion of the needle directly into the skin substance. It is therefore endermic and the skin is seen to become blanched as the needle advances delivering its solution on the way. But little of the fluid is pressed out as the needle advances. When the syringe is empty or the needle has advanced to the limit of its length, refill and insert just inside of the last blanched spot and proceed in a line until the end of the contemplated line of incision is reached.

Pressing out too much of the solution at one time causes a burning sensation and should therefore be avoided as the only pain should be that of the initial prick of the needle. Care, however, should be taken to inject just sufficient of the solution to penetrate beyond the zone of operation laterally, to insure sufficient space for the insertion of sutures into anesthetized tissues. Only a small quantity of fluid is necessary in this procedure as it comes in direct contact with nerve terminals. By touching the injected line with the needle in several places along its length and inquiring of the patient if it is felt, we may make sure of the completeness of the anesthesia before making the incision which should begin and end inside the anesthetized area.

Subdermic Method. An appreciable area of skin and subcutaneous tissue may be incised by anesthetizing as previously described, together with depositing the fluid well under the skin, thus affecting many terminal nerve branches before they reach their final distribution in the skin, and widening the anesthetized area considerably.

This method is applicable to such work as the removal of small growths, and the deep incision of a carbuncle. Beneath the skin in the loose connective tissue the fluid is deposited and causes anesthesia by acting upon the nerves just before their emergence into the skin. The two methods may be combined. It is not possible to inject directly into thin skin or mucous membrane and it is therefore employed in such operations as circumcision, where the nerve terminals must be anesthetized by the diffusion of the anesthetic from its position under the skin. A little time should be allowed before beginning the operation to permit of the diffusion of the drug. This applies also to such operations as that for ingrown toe-nail where the deeper tissues down to the root of the matrix are involved.

Edemitization Method. This is the method of Schleich and it is to him that the credit must be given for a procedure which has done more to encourage the use of local anesthetics in operative surgery than any other. He employed weak solutions of cocaine and other local anesthetics in great volumes of water in order to gain the combined action of both drug and of pressure. The method is described under the heading of "Cocaine." It was designed to obtain anesthesia with cocaine with the elimination of the toxic effects of the latter.

There are decided disadvantages to the filling up of the tissues with fluid; healing is delayed; relations are distorted and coaptation of the edges is difficult. This is probably the method of selection where an indefinite amount of manipulation is expected and where the length and depth of the incision may need to be augmented. A large quantity of a very weak solution is employed and the tissues in all directions are injected until visibly distended.

Nerve Blocking Method. By injecting a small quantity of a fairly strong anesthetic solution either directly into a nerve or beneath its sheath, the entire area supplied by it will be anesthetized. This method of nerve blocking may be spoken of as *endoneural* when the injection is made directly into the nerve trunk, and *perineural* when made into its sheath or immediately outside of the nerve. The injection of fluid around nerves too small to inject directly is also spoken of as perineural nerve blocking. (Hertzler).

DRUGS EMPLOYED

The essential qualities of a good local anesthetic are:

1. Reliability in producing anesthesia.

2. Constitutional and local harmlessness.

3. Non-irritating qualities.

4. Ability to be rendered aseptic by boiling.

No one local anesthetic can be exclusively relied upon to fulfill all of these requirements at all times. Each one has its advocates and from the large number offered, it is possible to select several which, while not being perfect, are preferable to cocaine in that they obviate the disagreeable train of symptoms peculiar to that drug.

By local anesthetics are understood certain chemical compounds, weak solutions of which, when brought in contact with sensory nerves paralyze them without lastingly injuring them. This effect is dependent upon the presence in these agents of certain atom groups which Ehrlich named *anesthiferous*. It is possible that just these atom groups enter into certain chemical combinations with the nerve substance and that the nerve thus remains paralyzed until the newly formed compounds are split up and the poison is washed away by the circulating blood.

Cocaine is the original type of a local anesthetic. Einhorn has made possible its synthetic production and has also opened the field for a great number of experiments of scientific and practical importance leading to the discovery of new local anesthetics obtained by exchanging the non-anesthiferous atom groups of cocaine for other groups different for each of the various new agents; thus eucaine, orthoform, anesthesine, alypin, and others have been obtained.

Cocaine occurs as a white, crystalline powder, readily soluble in water and in alcohol. It is an alkaloid which effects all living protoplasm. It first excites, then paralyzes. In greater concentrations it paralyzes immediately. Its effect is very ephemeral, producing no lasting harm to the cocainized protoplasm. Its effect is most readily understood by assuming that cocaine poisons the protoplasm by entering with it into combinations which are easily broken up. The products of decomposition, among which cocaine cannot be recovered, are slightly or not at all poisonous and are carried away by the circulation.

Effect on the Mucous Membrane. The external application of cocaine in solutions of varying strengths has been of great service since its introduction by Roller in 1884, and many operations on the eye and on its coverings are now greatly facilitated, by reason of its use. Small quantities only are required, hence there is little fear of its toxicity. Its anesthetic qualities by contact are also made use of in operations in and about the nose and throat. Here comparatively mild solutions are used liberally but care must be exercised against its noxious effects; it is usually employed in freshly prepared solutions which are held to be less toxic. Where extensive areas of mucous membranes are to be anesthetized, as in the rectum or urethra or bladder, one of the less toxic drugs is preferable.

Strength of Solutions. In the eye, it is customary to employ a 4 per cent. solution. For work in the nose, 2 per cent. is generally considered sufficient. In the latter connection, it is often combined with adrenalin solution in small amounts to mitigate its depressing effects as well as to control bleeding. The latter effect is but transient and is omitted by many as unsatisfactory because of the more profuse subsequent hemorrhage. In this respect cocaine and adrenalin are similar. They both cause constriction of the minute superficial vessels and immediate blanching of the membrane; work in the nose is hence greatly facilitated, the field of operation being clear and enlarged by the shrinkage of the encroaching membrane, but it is incumbent upon the operator to keep his patient under observation at least an hour after the completion of the operation that he may be certain of the degree of hemorrhage after the effects of the drugs have passed away. For the above reason many operators prefer a general anesthetic or one of the local anesthetic drugs which exert no constrictor action so that they may know, *ab initio*, the exact degree of bleeding.

Whatever drug is used, strong solutions are seldom necessary for application to the mucous membranes but the necessary time for its absorption is a prime requisite. To secure anesthesia of the conjunctiva and cornea, the solution is dropped into the eye at the outer canthus and as it flows off with the tears, it must be replenished three or four times until anesthesia is accomplished. In the nose, a spray over the site of incision or a pledget of cotton saturated with the anesthetic solution and allowed to rest in contact with that locality, will suffice. The flow of mucus from the nasal mucosa is stimulated by the presence of the cotton pledget and it soon becomes entirely coated with a thick mucus which no longer is able to impart to the membrane its anesthetic solution and must therefore be renewed several times before complete insensibility of the part is assured. The topical application of a strong solution on a cotton wound applicator to a limited area or spot is also efficient.

Application by Injection. In order to bring the anesthetic in contact with the nerves, it is necessary, where a skin surface is to be incised, to inject the solution as already described. The technic, previously detailed, applies here, and any of the methods may be employed for the injection of solutions of cocaine, some preferring a single method to the exclusion of all others. The locality to be treated will also influence the operator as to method.

Endermically. The endermic method is the one most generally employed in securing cocaine local anesthesia by injection. The papillary layer of the skin is well infiltrated with a mild solution (one-eighth per cent. to one-half per cent.), frequently with adrenalin 1-1000, in the proportion of 15 to 20 drops to the ounce of the solution. The strongest of the formulas of Schleich may also be used for endermic infiltration.

The skin is injected to a fair degree of tension and a white ridge marks the line of injection which should be sufficiently extensive to permit the manipulation of the cut edges.

Edemitization. Schleich's solutions are here of extreme value because large amounts of solution are necessary to produce the degree of distention required because of the minute quantity of cocaine present, though the added salt and morphine assist considerably.

Nerve Blocking and Perineural Blocking. Here a stronger solution must be employed; 1 per cent., or even stronger, is injected in small quantities, either into the substance of the nerve or under its sheath, as already described.

Strength of Solution. Schleich has worked out a method whereby very weak solutions of cocaine may be used advantageously. His plan is to enhance the action of the drug by the admixture of morphine in minute quantities and of sodium chloride in proper strength. These substances, in themselves, were found to possess anesthetic powers. Large quantities of Schleich's solutions may be injected—even several ounces, without ill effects as they contain so little cocaine. The formulas used by him are:

1.	Cocaine hydrochlorate	0. 2
	Morphine hydrochlorate	0. 02
	Sodium chloride	0. 2
	Distilled water	100.
2.	Cocaine hydrochlorate	0. 1
	Morphine	0. 02
	Sodium chloride	0. 2
	Distilled water	100.
3.	Cocaine hydrochlorate	0. 01
	Morphine	0. 005
	Sodium chloride	0. 02
	Distilled water	100.

It will be seen that the strength of cocaine in the respective solutions is from one-fifth to one-hundredth of a gram.

The solutions used in the early days of cocaine anesthesia were much stronger than were found necessary afterward and it has now become the rule to employ weak solutions and to give them time to penetrate the tissues. The less toxic action of mild solutions, even when like amounts of the drug are employed, makes it incumbent upon the operator to follow this plan and the element of time is so important in the matter of securing a perfect local anesthesia that it is customary to wait fifteen or twenty minutes after the completion of the injection before making the incision. The weakest solution possible is the one of choice in the use of this anesthesia.

Toxicology. The repeated use of cocaine in the same patient should be avoided on account of the danger of establishing the cocaine habit. The drug should be given with the greatest care, especially in operations about the head, neck, face, and urethra, as several deaths and many alarming cases of syncope, delirium and paralysis or tetanic fixation of the respiratory muscles have followed its use. Because of its marked depressing effect upon vital organs, it should never be given unless the patient is in the recumbent position. The administration of one drop of a one per cent. solution of trinitrin given at the first onset of the constitutional effects and repeated if necessary every five minutes, will entirely prevent any unpleasant effects as it is a true physiologic antidote.

If the surgeon has a case in which he intends to use large amounts of cocaine, it is best to have at hand and ready for use the following agents: a hypodermic and a rectal syringe, a battery, cardiac and respiratory stimulants, oxygen, and a catheter.

If the patient becomes very delirious and is in no way depressed, chloral or hyoscine should be given. In all cases of cocaine poisoning the patient should be catheterized to prevent re-absorption and should then be treated symptomatically.

Strong solutions should never be employed for any purpose except in cases where, by previous experience with the mild ones, it is known that no idiosyncrasy exists.

The central nervous system, and next the sensory and motor nerves, are affected by cocaine. Respiratory paralysis follows the introduction of appreciable amounts of cocaine into the circulation and respiratory depression may follow the introduction of smaller quantities. A given quantity of the drug in great dilution will, under normal conditions, give no toxic symptoms, whereas the use of the same amount in a more concentrated form will give rise to pallor, cyanosis and even syncope and collapse. It is said that a maximum dose of cocaine can never be fixed; this, however, seems of less importance than knowing the minimum dose, for while it is true that many bear it well, this drug so frequently gives rise to toxic symptoms, and the idiosyncrasy for it is so common, that one can never be certain of an exact dosage. Various pharmacopias place the maximum dose at 0.05 grm. (about seven-eights of a grain).

Bearing in mind that a great dilution of a given amount makes for safety, we are astonished to learn that 7 c. c. (about 2 drams), of a 1 per cent. solution introduced into the urethra has caused death. (Czerny).

Hertzler cites numerous instances in which a few drops of a more concentrated solution (2 per cent. to 4 per cent.) have caused death. It is therefore obvious that the use of this drug must be guarded by a technic so perfect that but the smallest quantity of a very weak solution shall be permitted to enter the circulation.

Adjuvants, Substitutes and Safeguards. The numerous disadvantages in the general use of this most efficient but most treacherous local anesthetic have operated so strongly that efforts have constantly been made to find a substance which, when used with it, would correct its toxic effects.

The desirability of employing large quantities of an anesthetic solution so as to enable the operator to infiltrate large areas of tissue has led to the method of preparing very dilute solutions and mixing them with various chemical substances which in themselves would act as mild anesthetics and at the same time increase the diffusibility of the cocaine. With any of these substances, cocaine still remains toxic and the quantity injected must be kept account of when an operation of any extent is being performed even though the solution be never so mild.

A valuable preventive to this absorption is found in the application of a constricting band or tourniquet to impede the return circulation and allow the washing out of much of the drug before the obstruction is removed. It is evident that no method has yet been devised whereby the use of cocaine is rendered safe and it is for this reason that chemists throughout the world have sought to produce either a new anesthetic drug or to evolve a drug synthetically, from cocaine, minus its toxicity. This has been done, but cocaine still has its adherents because of its superior qualities.

Quinine and urea hydrochloride is one of the new substitutes which has found much favor. Among the synthetic derivatives may be mentioned alypin, novocaine, stovaine, betaeucaine, tropacocaine, anesthesin, subcutin and many others. Each of these has its advocates and all of them have some advantage over cocaine; they have disadvantages as well, which, however, in the hands of skilled operators, may be overcome.

Quinine and Urea Hydrochloride. Among the quinine salts and combinations, the above has found most favor. It consists of a molecule of quinine hydrochloride and one of urea. It occurs as a fine crystalline powder and is readily soluble in water, forming an acid solution.

This substance is one of the most recent and best substitutes for cocaine, being capable of a wide range of usefulness and practically devoid of any toxicity. It causes redness on being injected and, in strong solutions, may delay healing considerably, this constituting the main disadvantage to its use. After the use of this anesthetic, primary union is not usual.

In a one per cent. solution, anesthesia is accomplished by any of the methods already described. Weaker solutions require a more perfect technic, and are therefore not generally employed. They, however, are indicated where it is imperative to secure primary union and when for some reason no other local anesthetic is available. The scar formation which almost always follows the use of this anesthetic would indicate that some other drug be employed in operations about the face and neck. This anesthetic is preferred by many because of its safety in large quantities and because of the length of insensibility following the injection of solutions of from 1 per cent. to 2 per cent. strength.

Notwithstanding knowledge of the facts above enumerated as to the difficulty of primary union and the likelihood of scar formation in connection with the use of urea and urea-hydrochloride for purposes of local anesthesia, this drug is still considered a most valuable and useful one for providing local anesthesia for operative purposes.

Novocaine. This drug is one-seventh as toxic as cocaine but is also weaker in action. It does not cause vascular constriction but has a preliminary vasodilator action. Like quinine, it has a decidedly irritating action when injected. It has a decidedly toxic effect when used in stronger solutions than 2 per cent. and causes tonic and clonic spasm. In a 1 per cent. solution it is probably safest and best as an anesthetic and one-half ounce of such a solution may be injected without fear of unpleasant consequences.

Its dose is said to be about seven grains, but this may often be the cause of alarming symptoms, and half of this quantity would perhaps be a safe limit. The duration of anesthesias of fairly strong solutions is about fifteen minutes; the action is more prolonged if used with adrenalin.

Various combinations of drugs besides adrenalin are employed with novocaine. Fischer recommends its use with thymol, but even so, it is not efficient for a longer period than twenty or twenty-five minutes.

Novocaine is frequently used in alcoholic solutions for injection in neuralgic subjects. The commercial tablet of novocaine and adrenalin is convenient for office use.

Alypin. This substance occurs as a crystalline powder, easily soluble in water, alcohol and ether, and makes a neutral solution.

Alypin is in every respect the equal of cocaine though not quite as strong. Schleich has found that its use, in conjunction with minute quantities of cocaine, permitted of a reduction of the entire amount of anesthetics necessary to accomplish insensibility.

In its use on mucous membranes it does not cause any anemia and therefore no secondary bleeding occurs. This is a great advantage also in the examination of mucous membrane lined cavities, such as the eye, nose, throat and urethra, inasmuch as after the application of cocaine, the blanching of the membrane conveys no idea of the real condition of the parts.

Because of the results he obtained, Schleich now recommends the following solutions for infiltration:

1. Cocaine 0. 1
   Alypin 0. 1
   Sodium chloride 0. 2
   Distilled water 100.
2. Cocaine 0. 05
   Alypin 0. 05
   Sodium chloride 0. 2

Distilled water 100.

3. Cocaine 0. 01

Alypin 0. 01

Sodium chloride 0. 2

Distilled water 100.

For other operative procedures of a minor character, it has been found that one-fourth per cent. to one-eighth per cent. is sufficient. For application to mucous membranes, as in the urethra, nose and throat, 1 per cent. to 2 per cent. has proved effective.

Stovaine. Stovaine is used more for spinal anesthesia than for local purposes; it is said to work well in inflamed tissues.

Several drugs have been used because of their lessened toxicity and many are constantly being tried but to be abandoned because of their inefficiency or irritating qualities. None of them are as efficient as cocaine and the weak solutions of Schleich are about as active as stronger solutions of many of these and are not more toxic.

Among the other cocaine substitutes in general use are betaeucaine, tropacocain, anesthesin, and subcutin.

These all find a special field of usefulness, but for general work, are limited, because of some disadvantages which each and all of them possess.

Individual selection plays an important part in the use of a local anesthetic, and one operator, by practical experience, may obtain results with a given drug, which another fails to achieve.

The essential feature to be remembered by the practising chiropodist is, that the use of any drug employed for anesthetizing purposes, even though but local, should be safeguarded in every way.

Cold. The methods of using ether, rhigolene, or ice and salt, to produce cold, are slow and unsatisfactory. If cold is to be used to produce local anesthesia the most efficient and convenient method of applying it is by means of *ethyl chloride*. This fluid is very volatile and is best controlled by having it in air-tight tubes. When not in use, a valve covering one end of the tube prevents leakage. When the valve is pressed upon, the orifice of the tube is opened and the heat of the hand forces out a fine stream of the liquid which is directed upon the parts to be frozen. Rapid evaporation causes intense cold. The nozzle should be held about fifteen inches from the area to be acted upon. When the spray strikes the integument, redness almost instantly results but in a few seconds the part becomes hard and white. This condition indicates local insensibility and lasts about two minutes. If the action is slow, it can be much hastened by gently blowing upon the parts to increase the rapidity of evaporation.

The refrigeration method of local anesthesia is of limited usefulness and is recommended only for the opening of felons and abscesses, for removing wens from the scalp and back, and for producing a painless area in which a puncture is to be made. It must be borne in mind that sloughing and ulceration of the skin are liable to follow the use of cold.

Work under this form of anesthesia must be done with rapidity not always consistent with thoroughness, and should therefore be employed only when a single incision or puncture is indicated.

The pain incident to subsequent thawing is severe and, in general, is about as hard to bear as an incision without an anesthetic.

For the purposes of practical podiatry, the chiropodist is advised to use a substitute for cocaine rather than the cocaine itself when local anesthesia is necessary. In the clinics of the School of Chiropody of New York, novocaine, quinine and urea hydrochloride, and alypin are preferred, and no single instance of toxemia has ever been experienced. There have been cases in which the anesthesia did not prove thoroughly effective, but, in the main, these drugs have well answered the purposes of their use.

THE END